T0254557

Hacking Health

David Putrino

Hacking Health

How to Make Money and Save Lives
in the HealthTech World

 Springer

David Putrino
Icahn School of Medicine at Mount Sinai
New York, NY
USA

ISBN 978-3-319-89089-0   ISBN 978-3-319-71619-0   (eBook)
https://doi.org/10.1007/978-3-319-71619-0

Printed on acid-free paper

This Springer imprint is published by Springer Nature
The registered company is Springer International Publishing AG
The registered company address is: Gewerbestrasse 11, 6330 Cham, Switzerland

To Dr. Jean F. Coppola, who taught me how, when, and why to use technology to enrich communities and impact public health. Your vision, passion and friendship will be greatly missed.

Acknowledgements

Writing any sort of book tends to be a fairly solitary and selfish endurance event. As such, I am incredibly fortunate to have an incredible support network of wonderful people who helped me through this process. Thanks to these people, I even managed to have a bit of fun writing this one, and I hope some of that comes through to you all. I am especially grateful to the following:

To my wife, Rose, for putting up with all of the complaining, late nights and early mornings with (mostly) good humor and lots of love.

To my parents and brothers for bringing me up in a loving, supportive, and enriched environment that has kept me curious, confident, and motivated over the years.

To Nichole, John, Bunny, David, Robin, Marta, Phil, and Rachel for patiently sifting through all of the disorganized thoughts that made up my first drafts, helping me to make this thing readable.

To Rusty, Brandy, and Brad for all of their amazing reading recommendations.

To Flux, for his amazing animations.

To our dog, Brasky, for taking me for daily walks.

Contents

About the Author

David Putrino is physical therapist with a Ph.D. in Neuroscience. He worked as a clinician in Australia, before moving to the USA to study computational neuroscience at Harvard Medical School, MIT, and NYU. He has served as a Faculty Member at Weill Cornell Medicine and Burke Medical Research Institute. He is currently the Director of Rehabilitation Innovation for the Mt Sinai Health System and an Assistant Professor of Rehabilitation Medicine at the Icahn School of Medicine at Mt Sinai. He works to develop innovative technology solutions for individuals in need of better healthcare accessibility. He consults with the Red Bull's High Performance division, using evidence-based technologies to study and improve athletic performance. In his spare time, he volunteers for Not Impossible Labs, a group that creates disruptive technological solutions for high-impact humanitarian problems. In addition to a number of academic publications, many of David's projects have been featured on ABC, Sports Illustrated, the Wall Street Journal, the BBC, Time Magazine, Wired Magazine, and the LA Times to name just a few. He lives in Brooklyn with his wife and dog.

Part I
Nuts and Bolts

"I was taught that the way of progress was neither swift nor easy"
                                                                —Marie Curie

I was eight years old, laid out flat on a big rug in the living room of my parents' home in Western Australia. I was watching *Star Trek IV* (circa 1986) with my brothers: It's the one where Captain Kirk takes his crew back in time so they can save the whales. I was completely enthralled…even at eight years of age, I was already a massive nerd. There's this fantastic scene in the movie: Pavel Chekov, the Navigator of the U.S.S Enterprise, has sustained a head injury and lies comatose in a 1986-era hospital in San Francisco. Chekov is in a lot of trouble—the surgeons in the hospital are looking to drill a hole in his skull so that they can relieve the pressure on his brain. He is at the mercy of twentieth-century medicine, and things are looking real bad. Just as the "primitive" surgeons are preparing to scramble our favorite navigator's brains, the hilariously sassy Doctor Leonard McCoy sweeps into the operating room majestically. He brutally excoriates the surgical team and pulls a couple of small devices out of his pocket—the first effortlessly diagnoses the problem: a burst blood vessel in the brain; the second heals the damage in a matter of moments. Chekov is fully recovered, McCoy takes a final jab at the barbarism of 1986's medical prowess, and they leave the operating room in triumph.

Fast forward more than 30 years on from that fateful scene in Star Trek, and I'm ready to admit that TV may have lied to us. Health technology is on the brink of a revolution that we sorely need, but we're yet to hear the first shots fired. Most healthcare experts acknowledge that the unprecedented rate of global aging will place overwhelming strain on health resources. Traditional models of healthcare delivery will prove inadequate for dealing with the sheer volume of individuals who will be in need of quality care. The healthcare industry is facing an enormous gap in its ability to deliver care to a large number of underserved individuals. Technology presents the field with an unprecedented ability to bridge this gap. Many different and fascinating fields have emerged in response to the realization that pairing technology and healthcare may be beneficial: health technology (HealthTech), biodesign, electronic health (eHealth), bioengineering, biotechnology, digital health (dHealth), mobile health (mHealth), telemedicine, and telehealth just to name a few.

To keep things simple, this book will use the term "HealthTech" from here on out. Of all of the terms/fields listed above, I feel that HealthTech is the one that is broad enough to encompass all of the flavors of products and entrepreneurs that this book will reach and help.

The arrival of the HealthTech field has also created a burgeoning marketplace: The current global value of this industry is estimated at $85 billion (as you read these words, this number has already grown), which will continue to grow to an estimated $300 billion by 2022. Big numbers, right? These numbers alone have inspired many people to drop everything and get involved in HealthTech start-ups. In addition, there's an altruistic component to all of this, which means that HealthTech is also attracting a host of entrepreneurs and investors who are prone to soul-searching and/or self-loathing about their unapologetically capitalist lifestyles. Many of these ethically conflicted, tortured souls are flocking to the HealthTech marketplace because they see the opportunity "do some good and help people in a way that is still good business." From my perspective, I welcome this influx of experienced entrepreneurs, because the field certainly needs the help of a group of people with strong business experience.

Now, onto the bad news: Despite the size of the marketplace, HealthTech start-ups experience a much higher failure rate than a typical tech start-up. According to a recent contribution to Forbes, only 2% of HealthTech start-ups actually reach the point of revenue or adoption. I think that it is important to really highlight this statistic: We're not even talking about commercial success here—we're simply talking about a point where a HealthTech product reaches a marketplace where it can be bought and sold. This should be a frightening and sobering statistic for anyone who is involved in a HealthTech start-up right now. **The primary reason for this prodigious failure rate relates to the misconception that creating a HealthTech product is the same as developing any other tech product.**

The purpose of this book is to guide readers through the steps and logistics that are unique to the development of a successful HealthTech product. We will cover the "basic anatomy" of a HealthTech team and discuss the essential core competencies needed for effective HealthTech product development, including where one can find individuals with these competencies. In addition, a discussion of ethics in the HealthTech arena is crucial to this book—not just medicolegal ethics, but also *business* ethics, so we will certainly spend some time on that topic. The importance of strong team dynamics within a HealthTech start-up will be explored, and how gaining a fundamental understanding of each team member's individual motivations is crucial to the success of a project. Finally, we will cover a series of case studies that document some of my personal interactions with many companies that are attempting to make the move into the HealthTech ecosystem. These case studies will document both successes and failures of large-, small-, and middle-sized ventures. The purpose of the case studies will be to deconstruct each project, and provide an overall summary of the particular type of HealthTech venture that each case study represented, detail what went right or wrong, and provide a recommendation for the sorts of professionals that may find each particular venture helpful.

Let's also take a moment to quickly cover what this book **is not.** This is NOT a "HealthTech MBA for Dummies." We aren't going to cover topics in any depth if they relate to securing funding for your company, Venture Capital funding, Angel Investors, grants, business development, financing, etc. You should have someone on your team who understands these issues—there is no doubt that they are essential for success. They won't be covered here, however, because they are not my core competencies—I'm going to stick with what I know and let other experts handle the rest.

I have attempted to give this book a bit of a narrative, so you certainly can read cover-to-cover if that is the way you like to do things, but each chapter is also self-contained, so if you want to skip to sections that interest you, you can do that and I promise no one will be offended. Lastly, this book has been written with the intent of being largely free of jargon. When it comes to working within an inter-disciplinary team, it quickly becomes the language of choice for the sorts of sci-entists, doctors, consultants, and other professionals who want you to believe that "you can't do what they do." The problem has become so bad that studies have shown that the general population will perceive a rude speaker who uses jargon to be **more intelligent** than a colleague who is polite and easy to understand[1]. Unequivocally, that nonsense needs to stop—that is the sort of backward thinking that leads people to make bad investments based on misguided perception. We'll have none of that here. Learning should be fun, engaging, and accessible. I hope this book lives up to that notion.

Reference

1. Amabile TM (1983) Brilliant but cruel: perceptions of negative evaluators. J Exp Soc Psychol 19(2):146–156

Chapter 1
Introduction—Why Me?

"An expert is a person who has made all the mistakes that can be made in a very narrow field."

—Niels Bohr.

I come by my interest in Healthcare honestly—since, as a child, I spent a lot of my time in the hospital.

I was born in a town called Perth, which is on the southwest coast of Australia (strike one against me, you know what they say about Australians: we're a nation founded on a penal colony…certainly not to be trusted!). I was born to two first-generation Italian immigrants, who moved to Australia with their families in the 50s. As a kid, I spent *a lot* of time in hospitals. I was basically the child that most parents pray they never have. If there was a way to grievously injure myself in a specific location: family BBQ, playground, library, church courtyard, etc., I would not only manage an injury, but manage one so spectacular that I would end up in the ER getting many, many stitches. All things considered, though, I think that all of this time in hospitals got me interested in healthcare and healthcare delivery.

After my hundredth visit to hospital, I decided I was interested in rehabilitation, so I made the decision to study physiotherapy (Physical Therapy for the US readers, so let's just shorten to PT from here on out) at a small school in Western Australia called Curtin University of Technology. It is a profession that I love, and to this day I am very glad that I have the knowledge that I gained as a result of working in the field. My first step into the professional PT world was as a private practitioner. I had a little practice in the middle of nowhere, and I treated people for all sorts of aches and pains, helped people to recover from surgery, ran hydrotherapy classes for the elderly, and Pilates classes for a demographic that seemed to be primarily mums and professional football players. I was the local PT and I **loved** it.

However, there were elements of the profession that were very frustrating to me. The biggest issue was that we didn't really quantify our effect. Let's delve into this concept in a little more detail, as I realize it can sound a bit nebulous. When a person enters a PT office with a set of symptoms, our first step is to catalogue those symptoms on the appropriate clinical scales: pain location and intensity, affected joint range of motion and muscle strength, and so on. The problem is that most of

© Springer International Publishing AG 2018
D. Putrino, *Hacking Health*, https://doi.org/10.1007/978-3-319-71619-0_1

our means of measurement are largely subjective. As a PT, the fastest and most efficient way for me to gather information is by asking my client questions, observing the way they move, or putting my hands on them to physically evaluate how their muscles or joints feel. Now, this isn't such a problem for my acute cases, i.e. people who come in with a strained lower back or a sore neck because they slept funny. On cases like these, I can work my magic and get them feeling better in no time. But for people with more complex injuries, the people that require long-term management, it becomes very difficult to use subjective outcome measures to meaningfully track progress over time. And that is the heart of quantifying effect: taking an intangible concept, such as pain, and objectively tracking the progress a patient is making towards relieving it.

I think that deep down, most PTs are sensitive to the fact that it is hard to measure the effect we're having on our clients. Consciously or unconsciously, PTs tend to cope with this issue in a few different ways. The most common response is to ignore the issue entirely—use the methods you were taught in college, follow the latest approved guidelines, help as many patients as possible, and try not to think about the more difficult cases that eventually stop visiting you because you aren't helping them. The next most common response is to over-correct. A subset of therapists become hyper-aware of the need for quantification and strive to remain up-to-date on every possible device on the market that could help to prove that their treatment was (or was not) effective. Before too long (and before they bankrupt themselves), this crowd usually ends up in either the field of high-performance sports, or research. These are worlds where even minor bumps in athletic performance and/or incremental scientific discoveries are incredibly important, which permits them a reasonable budget to support their cravings for all of the coolest new toys. A final group of therapists turn into what I like to call the "sales-therapists". This group is into any and every fad out there and their predominant strategy is to **give the patient what they want**—acupuncture? Sure! Manual therapy? Absolutely! Xhosan throat singing? I'll, uh, get my guitar (frantically Googles throat singing…).

It didn't take me long to realize that I belonged in the research world, and I knew that I was particularly interested in recovery of function following brain and spinal cord injury. I decided to complete a PhD studying the neuroscience of movement control. Under the patient guidance of my two advisors, Dr Soumya Ghosh and Dr Frank Mastaglia, it was the PhD process that taught me how to be a scientist—how to ask good questions (*pro tip: there are such things as stupid questions*), design experiments, analyze data, write papers, and communicate research findings. As I look back on my own PhD experience, and now the grad students I have supervised over the years, completing a PhD seems less about answering a couple of well-crafted scientific questions and more about learning *how* to learn. If I did gain any "super power" as they put that PhD graduate cap on my head (figuratively, because I actually skipped the graduation ceremony…), it was how to become well-versed in any field very quickly. As with all good research, by the time I had finished my PhD, I had answered one or two research questions with marginal success, and I had generated about five thousand more that had me itching to do more research. But before we talk more about research, let's talk a bit more about stuff that happened while I was doing my PhD.

Throughout this book, there are going to be several themes that I center on, because I think they're extremely important to any HealthTech product team. A theme that you'll see recurring is the critical need for a multidisciplinary approach to healthcare delivery. Coming from a rehabilitation/critical care background, I've always been pretty sold on this idea. While I was completing my PhD, I spent my weekends and some of my nights working on a respiratory medicine ward as a PT at a place called Sir Charles Gairdner Hospital. When you work in a hospital, nothing does as much good for your patients as working in an effective team environment.

Effective teamwork in a hospital is no joke. Medical malpractice is now regarded as the third-leading cause of death in the United States—beat out only by heart disease and cancer [1]. To put that into further context, it is estimated that medical malpractice may be responsible for anywhere from 210,000–400,000 preventable deaths every year, and researchers estimate that non-lethal, serious harm as a result of medical malpractice may be 10–20x more common than this [2]. This is horrific, shameful, and really makes you think twice about EVER entering a hospital unless absolutely necessary. Let's talk a little more about malpractice. With some exceptions, malpractice rarely happens because someone is straight-up incompetent. It's easy to assume that the offending party was incompetent after the fact. However, in my opinion, accidents and mistakes in hospitals happen due to overwhelmed and/or chronically sleep-deprived individuals making incorrect care decisions because:

- They felt too busy and overwhelmed to check with the rest of the team,
- Their team had poor dynamics and were communicating little or not at all
- Their team was not giving good/trustworthy information

Patients can live or die based on our ability to gather crucial information from our care teams. Working in this environment taught me a lot about teamwork. More specifically, I learned how to work effectively with a team that I didn't necessarily choose in order to bring the best outcomes to patients under our care. When you have a full patient caseload, and you're running from ward to ward to treat your patients, knowing whose information you can trust, versus whose information you have to fact-check, saves you an enormous amount of time. In addition, the reason that a multidisciplinary team is important for healthcare delivery is because every bit of information matters. I hate to be the bearer of bad news, but here it is: we don't yet know the best biomarkers for most conditions we treat. This is why hearing from many members of a care-team about subjective observations is crucial.

I can't tell you the number of times I've walked into a patient's room to find that their numbers look fine, they seem fine (at first glance), and I'm inclined to let them rest because they appear stable. Then a member of the nursing staff (who I trust: see above!) tells me: "you know, his numbers look good, but I think he's deteriorating because he's more grumpy/tired/quiet/nauseous/something," and so I give them some extra attention, and the subjective response to treatment is staggering. On many occasions when I walked away from a clinical encounter, I would ruminate on the notion that had I gone by the book, without taking into account the subjective insights of someone who was watching my patient 24/7, my patient would have deteriorated and ended up in a high-dependency unit, intensive care unit, or worse. Yet, I've also worked with so many clinicians who won't talk to the nursing or allied health staff at all: "just show me the numbers," and if the numbers look good, they move on. This sort of blind faith in potentially faulty biomarkers is *not science*—it is the opposite of science, and leads to bad clinical practice.

Anyhow, I could go on forever about my experiences as a clinician, but that is not really what I'm here to do. The main thing I want to focus on is that healthcare delivery (especially for complex cases) **must be multidisciplinary**.

Following the completion of my PhD, I was fortunate to accept a post-doctoral research fellowship with joint appointments Harvard, MIT and Massachusetts General Hospital. Under the leadership of Professor Emery Brown, I spent almost two years working with him and his team, learning about big data and statistics and how to apply these techniques to exciting scientific questions about the brain. This is also where I started learning about the importance and intricacies of good data science, even when you're chasing a highly specific question.

When I first arrived in Boston, I didn't know a soul, no one could understand my accent, and despite my PhD in neuroscience, compared with some of the braniacs at MIT, I was just a PT who knew how to use Excel pretty well…actually, not that well…Emery threw me into his lab of "neurostatisticians" who were all bona-fide geniuses who were incredibly influential in their fields, and they introduced me to the world of data science. I remember looking on in wonder as they took my data and started using it to make predictions about what the brain would do next or if my experimental animal would move accurately or fumble. It seemed like magic to me, and I was instantly entranced. In Emery's lab I was learning about the power of good data science—what happens when powerful algorithms are employed appropriately within a scientific framework. I started to learn a lot about these seemingly magical tools that could make predictions and models about the real world from a subset of sampled data.

This was a really exciting time for me because I was learning a lot from a bunch of super-smart people. As I was being exposed to the good stuff, though, I also had occasion to peek into the abyss and observe the "dark side" of data science. I didn't really mean for that to sound so dramatic, but this is an incredibly important lesson for all health technologists to learn sooner rather than later: blind data mining is not good. Blindly mining data (we really should rename it "data fishing") is something that you do when you're handed a bunch of data that you didn't collect, or you did collect, but you didn't control how it was collected and now you're just looking for *something*. If you're working on a healthcare product and someone says to you "Well, we've collected a lot of ambient, background data, and now we're going to hand it over to our data scientists. They'll definitely find something marketable." Run. Run and don't look back. In Emery's lab at MIT, I was working with a number of incredibly talented data scientists, on rigorously collected and controlled data, asking a highly specific question about the role of correlated neural firing in the completion of a movement task.

Even still, I had to watch my collaborators like a hawk, not because they weren't trustworthy, but because they weren't physiologists: I was handing them data that was recorded from the nervous system, but no one sat them down to explain what we know about the nervous system. This means that they could actively pursue a model of neural firing that didn't make any physiological sense, but their algorithm was telling them it was the most probable model for the data they had. This could go on for days/weeks/months if left unchecked, and if you check in on your data scientist after months on the job and tell them they're doing everything wrong because you didn't give them well-defined parameters to work with, you're probably going to need a new data scientist. In the end, the big takeaway from my time in Emery's lab was that to *rapidly* benefit from the seeming magic of data science, you have to **know your question** inside and out. You need to understand the environment that you're interacting with deeply, and have a good idea of what you're curious about. I underlined the *rapidly* because if you're working for Google or Apple or Elon Musk, and you just have endless datasets that you want to mine, and your bosses don't much care how long it takes, then lucky you! For those of you in startup-land, however, don't be seduced by the erroneous notion that data science will magically give you all the answers if you find the right algorithm. It's the kind of magic that come back and bite you on your ass. Hard.

After working with Emery, I moved to New York University to continue my neuroscience research. At this point, the clinician in me was getting very anxious to move away from more basic neuroscience and into more translational research (that is, research that aims to collect scientific data and use it in more of an actionable, practical way). So I joined a lab that was interested in developing a Brain Computer Interface (BCI). BCIs are clever devices that tend to get a lot of press. Their job is to provide a conduit between the brain and some sort of machinery on the outside of the body. Probably the most widely known version of a BCI is the Cochlear Implant —a device that records external sounds, and converts them to electrical impulses that can interpreted by the cochlea (a part of the inner ear) as sound in the brain. Pretty cool.

We were trying to create a motor BCI that would allow people who are paralyzed from a high-level spinal cord injury, or even an amputation, to gain control of robotic arms. This was an initiative that was funded by the U.S. Department of Defense, and was pitched to me as a very clinical endeavor: we would do the breakthrough research that would make this work possible in humans. Cure paralysis? Sure. Needless to say, I was quite excited. I got to work with a great team of hard-working experts from all walks of life: engineers, data scientists, programmers, even animators and motion capture gurus. It was a fun project that actually met with a significant level of success. However, this experience quite jarringly taught me the difference between translational research and product design. I had been quite caught up in the notion that we would be "curing paralysis", but, in the end, we had a BCI that worked *OK* at decoding movement intention from brain activity, so long as one had a 2–3 million dollar robotic arm, a $100K+ brain implant (and a daring neurosurgeon willing to implant it), a couple of super-charged (yet portable) computers, and a few highly trained computational neuroscientists following you around 24/7. In its current form, the technology we developed did not have the ingredients of an affordable and accessible piece of assistive technology. In fairness, though, that wasn't really the goal of this particular project. Our team was commissioned to solve a specific engineering problem. Our Department of Defense contractor asked us to complete specific functional deliverables, like wanting our BCI to be able to process x amount of information every second, and control y different joints simultaneously in real-time. We completed our assigned engineering deliverables, but my own, deeper, goal of developing an assistive device for people with spinal cord injury was not satisfied. This was an important lesson for me, because it made me realize that I wanted to become much more clinical with my research.

After my time at NYU, I accepted an appointment as the Director of Telemedicine at Burke Medical Research Institute, and an Assistant Professor in the Department of Rehabilitation Medicine at Weill-Cornell Medicine. My job in that role, simply put, was to find ways that we could use technology to improve people's health and quality of life.

After three great years at Burke, I moved to my current appointment as an Assistant Professor of Rehabilitation Medicine at the Icahn School of Medicine at Mount Sinai, and the Director of Rehabilitation Innovation for the Mount Sinai Health System in New York. It has been quite a dizzying change of scenery to move from a tiny research institute north of NYC to the largest health system in New York City, but the change has certainly been a positive one. I work in a department that is a multidisciplinary, highly collaborative mix of clinicians, traditional scientists, clinical researchers and innovators. My role in the department is to develop a "HealthTech accelerator" within the hospital. I scout promising HealthTech companies that have good ideas but need guidance and clinical validation. Mt Sinai provides the space, facilities, access to our clinical team, and patient populations and we co-develop products together with the HealthTech companies.

Currently, we engage in HealthTech projects that cover a wide breadth of experience levels, from people representing the youngest of startups, all the way to several of the largest companies in the world. We have had successes, and many, many failures. One thing is for certain: everyone, from startups to massive corporations, sees the opportunity for profit in the HealthTech market. However, it is still an extremely young field, and ventures into the HealthTech market will continue to have a higher-than-average failure rate until we better understand what is needed for success in this marketplace. I truly believe there is a tremendous amount of social good that can be done by integrating advanced technology into the healthcare space. With HealthTech, we have a unique opportunity to significantly enhance the quality of care that we can offer to people who are in critical need of healthcare solutions that are accessible, affordable and simple. This is why I am passionate about ensuring the success of the HealthTech movement, and having it reach a point of success as rapidly as possible.

References

1. James JT (2013) A new, evidence-based estimate of patient harms associated with hospital care. J Patient Saf 9(3):122–128
2. Makary MA, Daniel M (2016) Medical error-the third leading cause of death in the US. BMJ: Brit Med J (Online) 353

Chapter 2
Constructing Your Idea

"You have to start with the customer experience and work backwards to the technology"

—Steve Jobs.

All startups emerge from an initial idea. Every successful startup owner will be able to tell you, misty-eyed, about that one dinner, train ride, drinking session or quick coffee where the idea was born. That point where they recognized an unaddressed gap in the market, or a way to do something significantly cheaper than the competition. Sometimes getting there is easy, and sometimes getting there is hard, but almost every successful entrepreneur has that "eureka" moment that is part of his or her company's DNA.

As a scientist, this process is very organic, because the entire scientific process comes down to scouring the literature in a particular field for a gap in understanding, constructing a testable hypothesis around that gap, and then designing an experiment to prove that hypothesis. Sounds simple, right? Find an exploitable area of the market, and then exploit it. In many industries, it *is* simple, or at least relatively so. Healthcare is unfortunately not one of those industries. It can be a very difficult and delicate ecosystem to navigate. For every potential product market, there are often issues like the following:

- Policy is lagging far behind technology.
- Healthcare industry workers are too busy to learn new, unreimbursed procedures for patient care.
- Insurers are only interested in what you can prove in a targeted, high-cost population.
- Patients rarely pay for HealthTech solutions out-of-pocket.

This book will explore these ideas in more detail, but for now I will underscore with a very clear warning: **As HealthTech entrepreneurs, you are entering a tough field. Developing a HealthTech product is a much harder task than product development in almost any other tech market out there.**

Entrepreneurs that enter the HealthTech field for the sole purpose of turning a profit should really look elsewhere. There are easier ways to make money than in health technology. It may be a generic piece of wisdom, but it does hold true in this situation: if you want your product to be successful, you need to be deeply passionate about the problem you're solving. Most entrepreneurs in most industries have heard this advice before, but it holds strong in the HealthTech field. Healthcare is deeply personal, and the right product has the potential to transform lives. That product can't be developed unless the team developing it is fully immersed in the area of healthcare that they're looking to impact. To make matters more complicated, even a perfectly made HealthTech product needs more than good marketing to sell. Clinical research must be carefully-conducted in order to rigorously prove that the product can truly do what it claims. This is not an easy task and it does not happen overnight.

Anyway, that's enough negativity; let's talk about the best way to be successful. If you're reading this book, you're probably an entrepreneur thinking about starting a successful HealthTech company. Let's pull a concept out of a hat—a budding notion that will be additive or helpful to the healthcare world:

I Want to Develop an App that Helps People to Recover from a Total Knee Replacement

Solid concept. In 2010, 600,000 Total Knee Replacement (TKR) surgeries were performed in the US alone [1]. The rate at which individuals over the age of 45 are undergoing TKR surgery has doubled from 2000 to 2010 [2], and are projected to rise by 673% to 3,480,000 procedures by 2030 [1]. Post-surgical rehabilitation services could be better, insurance only pays for so much physical therapy, and regular outpatient therapy is a pain in the ass to coordinate when you have a sore knee. Technology can **definitely help** to manage this whole process more efficiently.

At a glance, this is a concept that probably has some legs (pun intended). In fact, people are probably working on this exact concept right now. The right entrepreneur could likely walk into the right Venture Capital firm and convince someone to fund this concept with a little effort and a shiny pitch deck. However, if it isn't developed more fully before starting out, this startup is going to face a high probability of failure.

Let's dive a little deeper, because we don't yet have a fully-formed *idea* here—we're still working with an early concept. The next step in the process of constructing the idea is to ask—"who would want to buy a product built around this concept?" HealthTech usually comes down to a few basic markets:

- Patient service
- Primary caregiver/care community (Spouse or close family member, neighbors, good friends)
- Care provider (Doctors, allied health team, hospital)
- Payer (Insurer, government)

- Research tool (Universities, Research labs)

Let's think about examples of how this concept could be applied to some of these different markets:

PATIENT
An exercise app that provides the user with a home-exercise program for post-TKR recovery

Possible features:
- Schedules exercise sessions
- Reminds the user to exercise
- Logs progress and compliance
- Sends data to your physical therapist

CAREGIVER
A "chore-sharing" app that links the caregiver to a network of individuals around them in a similar situation to reduce the chance of burnout

Possible features:
- Social media-style app to encourage social interaction
- Educational resources about the normal TKR recovery timeline
- Links to counseling services and home-help

HEALTH PROFESSIONAL
An app that allows the therapist to quickly and easily prescribe exercises, track recovery metrics, use predictive analytics to support clinical decisions

Possible features:
- Library of common TKR rehab exercises
- Ability to add custom exercises
- Use phone/tablet sensors to measure functional activities and range of motion
- Research alerts for advances in TKR rehab

INSURER
An app that tracks post-operative progress and allows the insurer to stratify high-risk and low-risk patients recovering from a TKR to allow more efficient care-calibration

Possible features:
- Exercise reminders for the patient
- Compliance monitoring for the clinician
- Ability for patient or clinician to schedule an appointment
- Symptom monitoring to identify complications

As you can see, with very little effort our concept has now spun-off into four different potential product ideas. Each one is quite different and provides value in a unique way. Now, I know that a single concept turning into four ideas sounds like good news, but don't get too excited, we haven't established that any of these are financially viable just yet. In addition, if you are a young startup, it is categorically a bad idea for you to pursue all four of these ideas at once. Each of these ideas is quite different and the implementation team you assemble will also look quite different based on which idea you ultimately decide to pursue. However, building your team will come later. For right now, let's go over a couple examples of how different a HealthTech product might look when you tweak the audience just slightly, since each potential market requires something a little different.

A Patient-Centered Product

In the blue box, we detailed out a rough idea for an application that will guide patients through their exercise regime post-TKR. At face value, this is certainly a strong idea that meets an unaddressed and growing need. Furthermore, if the product is well-designed and well-implemented, both patients and clinicians can definitely benefit from using a system of this nature. However, I've already started making concessions: "well-designed and well-implemented." These concessions are non-trivial, but they are achievable with the right team. If you're going to move on this idea, you will need an app development team (preferably one with a background in designing products for the elderly) working closely with a TKR rehabilitation expert.

Next, let's talk about intended audience. It's been established that this is an app we're marketing to patients directly. In that case, how do we get patients to use it? Patients already get rehabilitation post-TKR, so why should they pay for this app? This is where having a TKR rehabilitation domain expert on your team is crucial since they have the experience and expertise to help design a product that can help address an issue they know is missing from the current service infrastructure. Once you've decided what that unique service could be, your next step should be to prove your effect. Are you able to prove that your app enhances outcome when used as an adjunct to therapy? Are you prepared to do that? Do you have funds put aside for a randomized controlled trial to show that your app improves outcome? Do you have a strategic partnership in place with a rehabilitation institute that will run a clinical trial for you? Are they prepared to design the trial and publish their findings in a timely fashion?

I know I'm throwing out a lot of questions, but these are the things that should be going through your head as you begin to plan your new HealthTech product.

A Caregiver-Centered Product

When we talk about HealthTech services that need to happen, I think there is no greater underserved market than caregivers. As our population continues to age, the primary caregiver to an individual recovering from a TKR is usually, by default, that individual's spouse. This can be a serious issue when it means that most of these caregivers are both elderly and often times managing chronic conditions of their own. Recovering from a TKR is no picnic; you often need help getting up out of bed, going from sitting to standing, and walking is tough early on and sometimes you're quite unsteady on your feet. Upon discharge from a hospital, it is rare that a person recovering from a TKR does not need some form of assistance in order to carry out daily activities.

Considering that, quite often, the person thrust into the role of primary care giver is an individual's spouse—someone of similar age and, often, similar impairment—adequate and safe care delivery is a serious concern. There is definitely an unmet need here, so how can we use technology to fulfil this need? The most obvious idea that comes to mind is a sort of "Uber" for caregiving chores. A pool of pre-approved individuals within a community is made available to assist with particular caregiving duties that range from self-care to cooking. Perhaps this idea is not limited to TKR—perhaps this is applicable to many different situations in elder-care and thus your app gets a cut of further proceeds, and all the while you're helping people receive income for helping seniors who are in need. I actually love this idea, so if anyone ever pursues it, remember that you read it here first!

However, I think that it comes with its own set of obstacles and difficulties. For instance, what is your liability if someone has an accident and gets hurt? How do you screen your online helpers to ensure that no crime or elder-abuse takes place? Is there a large enough market? What proportion of seniors of this generation is ready for an Uber-style app (asked a different way, how many 65 + year olds are currently using Uber)? How do you negotiate price-points for different tasks in a way that is "worth it" for the virtual caregivers and still affordable for the beneficiaries? These are all very fair questions that must be considered if you are interested in pursuing this idea.

Now, I'm not here to list down a book full of business ideas that I wish I could start in HealthTech. However, I hope that from these two examples, you will see that even when the target populations are superficially similar, each idea needs to be carefully constructed based on a critical need in the space and a fairly deep understanding of your target population's needs. This means that you need to really, genuinely understand the field that you are looking to impact. I volunteer my time across a lot of startup "incubator" and "accelerator" programs. These are programs that recruit early-stage startups and help them to build potentially interesting HealthTech ideas into a business (incubator), or help a more mature startup to get funding and scale (accelerator). In these roles, I see a LOT of different HealthTech startups, working in a wide variety of healthcare domains, and I've developed an eye for red flags when it comes to HealthTech startups that may be moving in the wrong direction. Of course, one will never be able to predict with 100% accuracy who is going to fail and who is going to succeed, but here are three traits/styles that I've noticed carry a high probability of failure:

1. *The technology-forward, idea-absent entrepreneur*: This person has developed a device (or software) and somebody has told them "you know where that would be *amazing*? Healthcare." I don't know why they were told that and I wish I could have been there to smack that person on the nose with a rolled-up newspaper. But I wasn't. So now we have a starry-eyed entrepreneur with no healthcare experience, looking to apply their technology to health. To be clear, this isn't necessarily a disaster: if the technology is genuinely good, and they are astronomically lucky enough to be paired with an expert that sees a perfect use-case, they might be ok. But those are some long odds.

2. *The "friend of a doctor" entrepreneur*: This can be one of the toughest cases to deal with, because there can be a lot of emotion associated with it. The scariest thing about this kind of entrepreneur is that it can be any of us if we forget ourselves. This entrepreneur had an *amazing* dinner, or brunch, or beer (or 10) with a friend who is a doctor. They will tell you (or show you) the napkins that they drew on when they realized that they had just solved all of the world's problems with one simple app/device. Now, don't get me wrong—all ideas have to start somewhere and this isn't necessarily a bad place to start. However, the next question is crucial: How many other doctors have you spoken to about this idea? The sad truth of startup culture in general is that your friends, your partners, your moms (especially your moms), even your seed investors are going to tell you that you have a great idea. This is heart-warming, but it can create the worst kind of echo chamber if you aren't careful. What you really need to do is ask a focus group in your chosen HealthTech field if *they* think you have a good idea. Experts who are unaffiliated with your potential product and don't have anything to gain (or lose) from telling you that they like your idea are going to give you the most valuable advice. I know that this is not exactly novel, ground-breaking stuff, but it would appear that people need to hear it again and again—**get a focus group**.

3. *The "we'll find something" data scientist*: This person. Oh boy. No domain knowledge, no specific idea, no specific question, no service in mind, just easy access to data and an almost religious faith that, this time, their 'novel machine-learning algorithm' will find something of value. It doesn't matter how often senior data scientists warn us, there is something endlessly seductive about the concept that we can effortlessly capture data from people, and then some magical algorithm will provide us with a data product that will be worth billions to the right buyer. Feeling skeptical? You should be. Time and again, I have seen projects like this chew up immense amounts of time and money before everyone concludes that they wished they had put more care into collecting a cleaner, more targeted data set.

I think we're off to a wonderful start. I hope that as we near the end of this chapter, I've convinced you that careful, specific construction of your idea is the first important step to developing an incredible HealthTech product. I hope I have also convinced you of the importance of being incredibly thoughtful about whom you hope to benefit with your product, and the exact way in which you hope to benefit this population.

I want to finish with a checklist of sorts. If you can answer these 8 questions easily, and you can find a bunch of unaffiliated experts who happen to agree with your answers, AND you think you can assemble a team that can actually generate the product at a reasonable cost on a reasonable timeline, then I think your budding product has a chance of success:

(1) *What specific problem, in the world of healthcare, am I solving?*
(2) *Who will benefit from mainstream use of my product?*
(3) *Who will be willing to pay for it? (Not always the same answer as Question 2!)*
(4) *Can my product be patented, or protected in some other way?*
(5) *How do I prove that my product works?*
(6) *What sort of regulatory approval will my product need?*
(7) *How quickly can I develop my product and have it approved for clinical use?*
(8) *Is there an industry or profession that will oppose or be threatened by my product?*

Now that you have constructed a good HealthTech idea, let's talk about constructing the right team for your project.

References

1. National Center for Health Statistics (2015). Hospitalization for total knee replacement among inpatients aged 45 and over: United States, 2000–2010
2. Martin, G.M., Thornhill, T.S., Katz, J.N.: Total knee arthroplasty. UpToDate (2014)

Chapter 3
Forming Your Team

"I've always believed you hire character and train skill"
—Lori Greiner.

With few exceptions, successful startup efforts typically involve a well-formed team. In the interest of completeness, let's start with some general advice for team building that every founder of a HealthTech effort should consider. In addition to this brief rundown, I would strongly suggest that you look into some of the many amazing books focused on hiring that are available, and I've included some in my reading list at the end of the book. Now, before you skip this chapter because "you know how to hire," I would ask you to keep in mind that a recent study of employers has revealed that they consider 20–50% of their hires to be "bad hires".

When I refer to a "bad hire" or "bad employee," I'm broadly referring to someone who is either not as skilled as you had hoped or someone who does not play well with your other employees. Additionally, you need to look out for employees who may be unethical and/or have a terrible work attitude. I don't want to sound too dramatic, but the absolute reality is that a bad hire can mean the death of your company. When you're dealing with a skilled professional hire, the cost of replacing a bad hire has reach anywhere from fifty to several hundred percent of that employee's annual salary [1]. Meanwhile, way back in 2003, the US department of labor tells us that the average annual cost of a bad hire is 30% of that individual's first-year potential earnings [2]. The negativity and toxicity that a bad hire can bring to your workplace is both unpleasant and costly. Tony Hsieh, the CEO of Zappos, famously told the business world that he estimated that bad business hires had cost him $100 million dollars [3]. These days, he proactively offers new hires $2000 to quit within the first week of training if they aren't feeling the fit with the company not a bad day's work! With that in mind, let's look at what is generally considered to be the most important hiring advice out there.

1. *Do your research*

You're busy, eager to get started, and time is money. It is tempting to hire someone if their CV looks right and they are charming in an interview. Hiring without doing your due diligence is akin to gambling, and it is the most common

© Springer International Publishing AG 2018
D. Putrino, *Hacking Health*, https://doi.org/10.1007/978-3-319-71619-0_3

hiring mistake that is made. CVs are easy to fake, puff up, and pad with information that people will never take the time to verify. There are specific courses that train people to appear competent, confident and charming in an interview. This can very quickly create the sense that you've found the "perfect hire," and lead you to an impulsive decision that you will later regret. So, here are a few simple steps that you should never skip:

- Ask the candidate to describe the circumstances that resulted in them leaving their last position (or last few positions if they've been moving around a lot).
- Contact their former employer for a discussion about their performance in their old workplace.
- Contact their listed references.
- Introduce them to other members of your team and ask them for their opinions (your hiring decisions shouldn't be unilateral).
- Look at their social media accounts/internet presence—I know that this is kind of creepy, but how people behave on social media can often give you a better sense of who they are than their persona in an interview.

2. *Be clear about your needs and expectations*

This is another issue that can easily result in a bad fit for a project. No one really *wants* to be a bad employee, but sometimes an otherwise skilled employee can find himself or herself in a job that is not a good fit for them. This can happen when the employer isn't clear about the position they are hiring for. Give your prospective employee as much information as you possibly can about their role. If you have a detailed plan, lay it out for them, and let them know about specific deliverables and timeframes. If you don't have a detailed plan, but you have a general goal in mind, then ask them how they would solve the problems that your company is trying to solve. If you are doing very specialized work, it is not out of line for you to ask for them to complete a small project for you to prove their skillset. Your role in an interview is to get as much useful information about the candidate as possible. Being as transparent as possible in the interview process will allow you to gauge how passionate they are about the project and how suitable they are for the prospective position.

3. *Experience matters*

Experience matters in many different ways. First and foremost, obviously you want to prioritize hiring an applicant who has experience working in startups (and even more ideally has people from other startups that can vouch for them). Being in a startup can be very stressful, hard work and some people that are new to the space quickly learn that it is not for them. If possible, you want your team to be made up of startup veterans—preferably people who have experienced a failure or two in the past, and can identify when things are going awry.

This may sound like a no-brainer, but I'll say it anyhow: if you can hire someone that you have worked with successfully in the past, make the hire! Hiring is such a scary, crucial, uncertain, NERVE-WRACKING process—if you have a candidate that you have worked with in the past, is easy to work with, and can definitely get the job done, then don't hesitate.

4. *Make friends with your applicants*

If you are highly selective about whom you hire onto your team, as you should be, it is very likely that a number of excellent applicants won't make the cut because they aren't perfect for the position. Good employers work to cultivate a pipeline of talent that they can call upon when they need new positions to be filled. A first step in cultivating this pipeline is ensuring that talented applicants who didn't quite make the cut are not just "dropped" when the best applicant is chosen, but are dealt with respectfully and courteously. In addition, the startup world is small and you never know when you might be reaching out to someone who didn't make the cut in your company for a favor at another company. Make sure that on that day, your former applicant remembers you as the nice employer that was timely, transparent and kind, rather than that jerk who never called them back!

5. *Avoid hiring toxic employees*

Although most of this section is about who to hire, these last two points will really be targeted toward who *not* to hire. Number one on this list is the toxic employee. Toxic employees can be extremely dangerous to your project *and* your workplace, and must be avoided at all costs. Toxic employees come in all shapes and sizes, but their effects on a workplace are always the same: they are frequently disruptive to the projects at hand, they are constant distractions to other employees, and they have a near-mystical ability to drain the energy out of the whole office. If you think back, at one point or another, you have probably worked with some form of toxic employee. Interestingly, researchers who study toxic workers have shown that the savings associated with avoiding a toxic worker are 235–640% greater than making a "superstar" hire [4]. To make matters worse, toxic employees are notoriously difficult to identify—it is a mistake to think that you wouldn't like a toxic worker during an interview, because they are experts at appearing charming and skilled in an interview. Dylan Minor, an expert on toxic workers and work environments, tends to boil the art of identifying a toxic worker down to the following four features:

- Overconfident: This means that they both overstate their abilities and often underestimate their chances of getting caught for unethical behavior.
- Selfish: Toxic workers are less likely to help a struggling coworker, and are more likely to take full credit for team projects. Here's an interview tip: ask them about team projects they have completed in the past and take note of whether they use "I" or "we" when discussing the team's accomplishments.
- Preaching strict adherence to rules: A toxic worker will tell you how fantastic they are at always following the rules. However, research into toxic workers show that although they boast loudest about following the rules, they are the most likely to break them.

- They are a bad fit for the position: Most of us have potential for toxic qualities to emerge if we're placed in the wrong environment. I brought it up before, but I'll say it again: curate your job description carefully, and hire someone who is appropriate for the job. Otherwise, they can quickly become toxic.

6. *Don't hire a team made up of different versions of you*

Look—you bought my book, so I like you…and I really hope that you like yourself. However, self-love is no reason to go and fill your company with multiple versions of yourself. Hiring a team is going to be one of the most crucial tasks that you complete in this process, and you need to make sure that your team is not going to agree with you all of the time just to be agreeable. NOTE: I'm not encouraging you to hire a pack of assholes that will be perpetually disagreeable! I'm just taking a moment to gently remind you that echo chambers crush both creativity and objectivity. You want team members that have skills and opinions that are different, but complementary to your own. This will require a little bit of navel-gazing on your behalf. Think about your technical and theoretical strengths and weaknesses and make sure you hire people with complementary strengths to your weaknesses.

A final point on exploring the basics: as an employer, you would be advised to embrace new technologies that assist startups in finding good talent. Online talent markets have begun to revolutionize the hiring process in the startup world. LinkedIn was a great start, but we've moved forward at a blinding rate in what social media can do for hiring. At the time of writing this, online organizations such as Freelancer.com, FounderDating and AngelList have made several hundred thousand introductions for entrepreneurs looking to become involved in startups. In addition, talent markets are beginning to emerge with the specific intention of accelerating health startups. These services make a lot of sense. McKinsey & Company tells us that using online talent platforms has the potential to increase company revenue by up to 9%, while reducing HR costs by up to 7% [5]. In addition, these talent platforms can effortlessly amplify your search from local to global, deepening your prospective talent pool, while boosting the global GDP by identifying and employing promising recruits in places you wouldn't think to look [6]. You haven't even started your business and you're already helping save the world!

Now that we've discussed some hiring basics, it is my duty to inform you that things get a little more complicated when you're dealing with a HealthTech product. I know that, once again, I'm telling you something you hear a lot: teamwork is important. However, the message I am trying to get across is that teamwork, and a multidisciplinary approach, are infinitely more important when you're designing a HealthTech product. This is because successfully building HealthTech products requires an unusually deep set of multidisciplinary core competencies when compared with other types of technology products. We are going to cover, in great detail, some of the most common professional types that comprise a Health Tech team later in this book, but for now, let's examine the role teamwork plays in HealthTech projects compared with regular projects.

Let's take a scientific experiment as an example. You want to run an experiment that will answer a specific scientific question. So, you pull together a team of people who have specific expertise within the field where your question lies. Typically, this team will be made up exclusively of scientists or scientists-in-training. The team has to get together and plan out an experimental design, conduct an experiment(s), analyze the data, synthesize the results and communicate the findings. If team dynamics are good, then everyone understands their role in the project, works together in a constructive manner, and all of the work happens in a relatively smooth, efficient fashion. If team dynamics are bad, the entire process is an ego-driven dumpster fire, nothing gets done on schedule, everyone leaves the project hating each other, and everything is awful. Yes—teamwork is important as most jobs/projects require some degree of it for you to get anywhere, but on the flipside, most jobs encourage teams to form that have common competencies and complementary skillsets. Things are a little different in the HealthTech ecosystem...

The biggest difference you will find is that your team is going to need members who have wildly different competencies to one another. Following on from this, you have to contend with the fact that HealthTech is an emerging field and historically, many of the professions you will be merging into a team have no experience working together. Now, I know that there are a percentage of you reading these words and thinking, *"well, I've designed digital/consumer electronic products before—I understand the need for a multidisciplinary team, this guy is just overcomplicating the matter"*. So let's dive a little deeper into this using our Total Knee Replacement (TKR) rehabilitation product from Chap. 2. To build this product out well, at the very least, your ideal product design team should consist of:

1. An allied health professional with strong expertise in TKR rehabilitation (Doctor, Physical Therapist, Occupational Therapist, Nurse).
2. A good software developer.
3. A user interface expert with experience in developing products for seniors and clinicians.
4. A regulatory expert.
5. A clinical researcher (preferably with experience in researching TKR recovery).
6. A patient advocate, i.e. someone who has been through the process as either a TKR recipient, or the primary caregiver of one.

Some of these team members may be paid employees or contractors, others may be part of your advisory board, some may hold equity in the company, and some may not. Regardless of the corporate organizational arrangements, you need skilled individuals to fill all of these roles, but you also need them to be able to work in a team environment. Now, a common first instinct for this type of project is to be incredibly linear by assigning clearly defined roles and setting a timeline to complete the project. I cannot be clearer: **this approach is a set up for failure**—as I said in earlier in this chapter, you are going to be bringing together professionals who don't usually work together. That makes you an innovator (isn't that cool?), but only if you work to maintain tight relationships between the people you're bringing together. Major project goals should be set in a way that involves vetting/opinion gathering from all members of the team before you progress to the next milestone.

I know all you cynics out there will tell me that I'm giving you a formula for designing a 'camel' (i.e. a horse that has been designed by a committee). I don't agree with this—I'm not saying that you shouldn't have clear deliverables and concrete product goals that you stick with, but if you are going to form a team that consists of a diverse group of professionals, you should:

- Solicit opinions from the individuals who will be directly interacting with your product (end-users) as early as possible, and then as regularly as possible.
- Have your product-builders watch your end-users interacting with the product.
- Have frequent team meetings to discuss user feedback and prioritize edits to the product.
- **LISTEN** to your users—if they consistently don't like something, address it: don't assume that you know better.
- Have your full team occupying the same physical space as often as possible.

Creating this manner of collaborative environment is truly difficult to do well. Our tendency is to let people work on their specialties in an uninterrupted fashion, but it is easy for members of your team to make a decision to "shift the goal-posts" if you don't have your professionals talking on a regular basis. For instance, your software programmer might decide that achieving a certain deliverable is going to take too long, so they create a workaround, sacrificing a crucial element of system usability rather than falling behind. A decision like this is something that must be discussed by the team, which ideally includes product end-users. Is the proposed hit to product usability worth the shorter development time? Or would it be better to stick with the planned system even though your development timeline will be lengthened? These are rarely easy decisions, but they should never be made alone.

A major goal of this book is to help you understand some of the usual suspects who get involved in HealthTech ventures a little better. So, let's take a break from talking about team dynamics and discuss some of the core members of a HealthTech team in more detail.

References

1. Gusdorf ML (2008) Recruitment and selection: hiring the right person. Society for Human Resource Management, USA
2. https://wdr.doleta.gov/opr/fulltext/99-testassess.pdf
3. https://www.youtube.com/watch?v=9C36EYM-mWQ
4. Housman M, Minor D (2015) Toxic workers.
5. http://www.mckinsey.com/industries/high-tech/our-insights/managing-talent-in-a-digital-age
6. http://www.mckinsey.com/global-themes/employment-and-growth/connecting-talent-with-opportunity-in-the-digital-age

Chapter 4
The Usual Suspects: Members of Your HealthTech Team

"None of us is as smart as all of us"

—Kenneth H. Blanchard

By now, we've established a couple of things that are of critical importance to your product: identifying a detailed product idea and constructing a solid team. Our next step is to talk about the sorts of people that make up a typical HealthTech team. Even the most experienced and skilled team can rapidly become toxic if there is not clear understanding, open communication and mutual respect of the different roles on the team. In this chapter, we're going to go through some of the more typical roles within a HealthTech team. Let's start by discussing what to aim for when building a HealthTech product development dream team:

- Chief Executive Officer/Leader.
- Engineer (software and/or hardware).
- End-User (Clinician/Patient/Caregiver/Insurer).
- Designer (User Interface/Experience expert and/or Industrial Designer).
- Regulatory guru.
- Scientist (preferably a clinical researcher with expertise in your area of focus).

Now, if you don't have a team like this lined up, or if you don't think you can arrange/afford to assemble this team at your current stage, don't panic—I called it a dream team for a reason! However, do think of this as your starting set of hiring goals. Every single one of these team members is incredibly important for developing a sound HealthTech product. As with every piece of advice in this book, following it *improves your probability of success* according to available evidence. However, if you can't assemble a team like this at first, it doesn't mean you're not going to succeed. In addition, just because I've listed six or so professions/areas of competency here, doesn't mean that you need to rush out and hire six or so people. Some of these people will be advisers, not hires, and some people can (and will) wear multiple hats. In many of my own projects, I have leveraged my experience as a clinician AND a scientist to help push a project forward. This is not always optimal, however. Striking a balance is crucial, and the more roles you ask your team members to take on, the less detail-oriented they will become out of necessity.

© Springer International Publishing AG 2018
D. Putrino, *Hacking Health*, https://doi.org/10.1007/978-3-319-71619-0_4

In short, aim to fill these roles in any way that you can, with the understanding that straying from hiring dedicated individuals for each role may end up making life a little harder…but that's OK, since you're an entrepreneur—you practically have "make life hard for me" tattooed on your forehead.

Getting down to business, the main purpose of this chapter is to give you a rundown of the crucial members of your HealthTech team: what they do, how they do it, and how they tick. A successful HealthTech product team is a HealthTech product team that communicates and respects one another. The first step in creating an ecosystem where open communication and respect is ingrained in the culture of your startup is making an authentic effort to understand everyone's process and background. Let's take a deeper dive working to understand each member of the team.

Chief Executive Officer (CEO)

The CEO is arguably the most important and least important member of your team at once. In business literature, you will often hear of CEOs being likened to orchestral conductors because no one knows exactly what they are doing, but when they do it well, you can really tell the difference. While the team is focusing on their specific parts of developing the product, the CEO is taking in the big picture by making sure the team is working effectively together and remaining motivated, ensuring the right people are talking/working together at the right times, fretting that the company's core mission is being fulfilled, and managing to keep the company above water from a financial point-of-view. It is a hard, often stressful juggling act. We aren't going to spend too much time going into the qualities of a good CEO in this book—there are already many great books out there to teach you these sorts of skills and we want to talk about things that a specific to a HealthTech product. However, I will impart some general advice, because I know that at least some readers are going to be first-time entrepreneurs and founders. Firstly, understand that being a CEO is a **real skill**: all too often it is treated as the meaningless title that comes hand-in-hand with being the founder of a company. If you are founding a company for the first time, you know nothing about being a CEO, and you're set on being the CEO of your own company, then:

(a) For the sake of all that is holy, READ everything you can get your hands on.

(b) Locate a good mentor and stalk (please don't actually stalk anybody) them until they agree to be your mentor.

(c) Bring on a fantastic advisory board for your company that can help you through your first time as a CEO.

If you have the funds (or a shiny enough product) to place you in a position where you can bring on an experienced CEO, then everything we talked about in Chap. 3 applies here, but most importantly:

(a) Prior experience in HealthTech (or healthcare in general) product development is preferred.

(b) Look out for toxic individuals (everything becomes way, *way* worse when they are at the top of a company).

(c) **Do your research**: what were the last companies this CEO worked with? Did the companies do well while they were there? Are they still doing well or is this a "pump and dump"-style CEO that boosts a company's numbers but then leaves before the artificially boosted numbers come crashing down? What do the employees at this CEO's previous companies think of them? This business is your child—don't hand over the responsibility of raising your child to an irresponsible, toxic narcissist.

I'm a scientist and a clinician by trade, so talking about CEO stuff often feels a little foreign to me. Fortunately, I'm blessed with a wide network of really talented friends who I can ask to share knowledge about these things. One such friend, Mr Brad Rinklin, is an absolute veteran executive for massive tech companies, and has worked under some very famous CEOs. I quizzed him about what he thought makes a good CEO and if he could recommend some good books for my reading list. He had the following advice to impart:

Most 'How to be a CEO' books are crap. I've read a few and find them either too basic or too self-serving to the egomaniacs that wrote them (or in many cases, commissioned them to be written).

I think being a CEO takes several key skills that you can develop, and continue to strengthen. Then, of course, there are the 'intangibles' like emotional intelligence, good social skills, etc—stuff that is inherent in natural-born stud-muffins* like you and I [*Authors Note: I have no idea what being a 'stud-muffin' entails…must be a Texan thing…]

The common traits I admired the most and felt like drove their success were simple on paper:

Sell the vision: From their making direct executive reports down to interactions with the front-line employees, good CEOs constantly evangelize the vision they have for the company, and the importance of key execution milestones. Then they regularly and clearly communicate wins and setbacks on the way to achieving the vision.

Listen: A good CEO always takes the time to listen to ideas from any source: employees, analysts, advisors, peers, and the board. I never see a good CEO get defensive when the feedback is critical or constructive. They treat all of these interactions as important "data gathering." The ideas aren't always good ideas but no matter what, a good CEO takes something away from every conversation…even if it is "OK, that person isn't as smart as I thought." Or, "that was a great idea that could really help us and it came from a recent college hire! I need to consider that person for other projects.'

Prepare for all scenarios: All good CEOs have back-up plans to their back-up plans. We had 3-year business plans that had milestones which would trigger slow-downs in spending or ramp-ups in investment depending on growth or product releases. These plans help us to separate uncontrollable market conditions from lack of execution on our end.

Hire the best and make them better: This one is probably the toughest. But I always observed that a good CEO searched for the absolute best talent, and will not settle for mediocre. Then, once they have these high-powered guns on board, they push the hell out of them—never giving them the chance to under-deliver. Every conversation centered on the same theme: "OK—you did a good job, but what could you have done better?" It is tiring sometimes, but great leaders make great people better. Push them higher than they thought they could go.

While that is a stellar rundown from Brad, if you are interested in reading more, I've added some good books on leadership (all cleared by Mr Rinklin himself) to the Reading List.

Engineer

Startups come in all shapes and sizes, so when I talk about your "engineer," I might be talking about a hardware engineer, software engineer, or your mate Doug from down the road who smokes a lot, but has always been good at building things. This member of your team, your builder, is what I'm going to be calling your engineer. Most people who fit into this category tend to share some personality traits that can either be incredibly detrimental or helpful to your product development. The most important thing to remember about a good engineer is that they will always build you EXACTLY what you ask for. My favorite character analogy for a good engineer is that they are a lot like the genie in one of those dystopian stories. You know, where someone makes a careless wish, failing to realize that the ensuing chain reaction places the planet into an awful spiral that causes a zombie apocalypse/dinosaur revival/Nickelback reunion tour or some other sort of extinction-level event. I'm willing to admit to being slightly dramatic here, since the biggest engineering horror stories that I've run into in the HealthTech market typically fall into one of two categories:

1. You get exactly what you asked for—a product that is capable of a truly amazing task—but you need one (or several) advanced degrees to understand how to use it.

2. You get exactly what you asked for, but it doesn't actually do what you had in mind. Now you have to sell a product that is slightly different, or less useful to your initial vision.

I really want to get the point across that engineers *build*, which means, you ask for it, you get it. Be HIGHLY SPECIFIC about what you ask for, otherwise you will be sorry. It should come as no surprise to you that the best way to avoid your very own product-extinction event is good team work. In the early stages of prototype development, you want your engineer in close communication with your designer regulatory experts and end-user advisers. Facilitating communication between these members of your team will ensure the engineer clearly understands:

1. The clinical goals of the product.
2. Basic design imperatives that will become important down the track when you move beyond your minimal viable product/basic prototype.
3. Honest feedback about usability necessities. Even if a HealthTech product solves a common healthcare problem well, it is much harder to have that product widely adopted if the end-user has go through a lengthy training process, or spend more time per patient. This is where honest and direct feedback from your end-users is absolutely crucial, otherwise you will end up with a product that works as advertised, but is so difficult to use that no one is interested in using it.

If you take anything away from this section (other than my 'genie' analogy), please let it be that giving your engineer a list of requirements, stepping out of the way to let them work, and then releasing whatever product they build to your end-users can be incredibly damaging to your product. Team communication and input at every stage of your build is crucial. This may sound super obvious, but never underestimate your ability to forget this advice when you're rushing to get your product out to a test population. Also, flip over to Chap. 5 for an example of how a large company I worked with got this all wrong.

End-User

I sit through a lot of HealthTech pitches. Many of them start in exactly the same way: "Recently I have had to interact with the healthcare system, <insert touching personal anecdote>, and this led me to realize that it is inefficient and garbage. Whilst navigating said garbage system I thought to myself, imagine if <insert product setup statement>, and that is exactly why I have created <insert snappy product name>."

Now, there is nothing wrong with this style of narrative being the impetus for you starting a company and building a product. We often get our passion from personally observing problems, it is laudable to solve problems encountered in a healthcare environment firsthand, and so many innovations have happened when someone from an outside field looks in and says "I can think of a better way to do this."

However, let me tell you where these pitches go horribly wrong: if this is the end of your narrative, you're probably in trouble. People with a strong pitch are the ones who go on to say something like "I then interviewed 1000 other patients with my condition and they said they would pay up to $x for this service," or "I discussed it with 50 doctors, and they estimate that this product would save them 10 min per patient," or "the CEO of <insert large insurance company> thinks that this will save them $y-million per year." These statements are the ones that will get the right people sitting bolt-upright during a pitch, because it means that you've isolated a problem that a specific consumer-base thinks is worth solving or will pay to solve. This is also why it is particularly important to have a product-adviser on your team that represents your target population. As always, the right person here is crucial: bring on someone who has been around the block a couple of times. If the product is for a patient or caregiver, you want to favor an experienced patient or caregiver advocate rather than a one-time patient. Experienced advocates have hundreds if not thousands of diverse cases and can tell you the value of your product across the board, rather than speculating from just a single, personal experience. If you're working with clinicians or insurers, favor feedback from experienced, senior professionals in your target market, the crankier the better, these are the people who make purchasing decisions for hospitals and see shiny new technologies come and go.

If you want to create energy around your product, hand it out to people in junior positions and let them play with it and get engaged and excited. BUT, make sure you do that only after you have designed a product that has impressed the grumpiest, most change-resistant and highly skeptical end-user you can find. Once you have found this person, throw them in a room with your engineer, regulatory expert and designer and begin the process of understanding what is actually able to be built (engineer), how simple and pleasant the user experience will be (designer), and how valuable such a product would be to the end-user.

I know that a lot of this sounds like basic market research (and it is), but it is worth mentioning once again that the sticky issue that differentiates a HealthTech product from others is the highly regulatory nature of the field. This is why you need to dialogue closely with experienced end-users—there are so many inefficiencies that exist in the healthcare systems all over the world, not because they are hard to solve, but because there is some silly policy, practice or pain-point that blocks any product from entering the space. These policies are often non-obvious, which is why experienced end-users are crucial to your success. Similarly, just because you can make a process better, doesn't mean that it will be viewed as valuable to the end-user if it adds effort. I see this happen a lot with tech products that try to target senior end-users. Sure, the product works, and sometimes will even solve a really important problem, but if it is too much bother to use, forget it—an example covered in Chap. 5 will touch on this in some detail.

Designer

Building a product that works is only half the battle. For commercial success, you also need to ensure that your product is highly usable, attractive, and inviting. A great product, in any arena, is one that the user actually enjoys using. When I talk about a "designer" in this chapter, it really depends on what your product is. If you have a software product, you have to hunt down some amazing user experience/user interface (UX/UI) professionals. If we're talking about hardware, find yourself an industrial designer. The importance of good design cannot be understated. To an outsider, the process can feel frustrating and intangible at times, but good designers are often as data-driven and evidence-based as scientists in their process, but with an added, artistic flair.

There is no doubt that good design can be the difference between a successful product and a total bomb. Still, so many people are unwilling to invest the necessary time, effort and money to ensure a nicely designed product. The obvious examples of design being a huge brand differentiator apply here as much as they apply in any other field—Google differentiating itself as a search engine, Apple taking the computing world by storm by making personal computing accessible to non-expert consumers. These brands achieved international success because they invested in design and made their products a pleasure to use. Doing so allowed them to leave their competitors, who were selling very similar products, in the dust.

I'm often blown away by how important good design can be to consumers. One example, which really sticks with me, relates to an ongoing project wherein my team and I, in collaboration with a company, have developed a device that can help to reduce the severity of certain movement disorders in individuals with Parkinson's Disease. We were trying out an early prototype on our very first user and the results were profound: his tremor was significantly improved, and he could complete a lot of functional tasks with his hands that were previously not possible for him. He was incredibly encouraged by the device and became quite emotional about the potential improvement to his quality of life.

When our first pilot test was over, we gave him the option of keeping the device (we had plenty of others) so that he could keep using it as we moved into the next stage of design. To our surprise, he politely declined, and his wife (and primary caregiver) declined a lot more vehemently. When we asked why, they told us that the device setup was too complicated and they didn't feel like learning how to do it. This is a totally fair statement that we're all familiar with—if a product is hard and annoying to use, your end-user won't use it. What struck me about this scenario, though, was that this device was so incredibly effective that our first user had been moved to tears. Yet, when offered the prototype for free, he was still completely unwilling to commit to daily use of the device until we had improved its design. Usability is such a crucial issue in product development. If you're developing something that you expect people to use on a regular basis, ensure that your end-user believes that your device is simple and pleasant to use (and be aware that this may be very different from what your perception of "simple and pleasant to use" is!). If you don't take care of this basic design imperative, your end-users are not likely to use your device, regardless of its efficacy.

Of course, design is always a balancing act where one must be careful not to prioritize style over substance. You want your product to look cool and be easy to use, but not at the cost of too much functionality. Furthermore, many designers have the heart of an artist and this means there is always the temptation to design for other designers. Meaning, they will make you a beautiful, impeccably designed product that every other designer will tell you is flawless, but your end-user will be lukewarm about it, because it wasn't designed with them, specifically, in mind. For this reason, your designer needs to spend a lot of time in a room with the end-user and the engineer during the early product design phases. There is typically some healthy argument to be had during the process where the team takes a product blueprint, funnels it through the designer's process (which may sacrifice some functionality to make it pleasant to look at and to use), and finally runs it by the end-user to see if it is something they will use on a regular basis. This is fine, a healthy argument is good, and you want your CEO to be checking in on the process closely in order to be the tie-breaker if the team reaches an impasse. This process will be non-linear, highly iterative, and should be as **transparent as possible**. What I mean by transparent is: don't let the designer or the engineer squirrel themselves away for months planning something that no one else can see. This happens a lot, and the person who goes on the missing list will always use the same excuse: *"well, I can't really show you what it looks like until it's finished" *waves hand at a computer screen of code, or a half-built prototype.** DO NOT FALL FOR THIS! If you're hearing someone say these words, it means they're building something, it isn't working out, and rather than tell you it isn't working out, they're improvising a solution without the input of the team. Too many monstrosities have emerged from basements with a manic, overzealous engineer or designer convinced they've created the perfect product in isolation. Things rarely go well from here.

There are a lot of great books that delve into the role of design in creating products that aren't just usable, but are actually enjoyable (even a little addictive) to use. I've added some to the reading list.

Scientist

You have an idea, you've made a prototype, now you must quantify your effect. I cannot state this often enough: **The critical need for quantitative, scientific validation is what separates a HealthTech product from any other tech product**. When you build a tech product, you can be as shady as you want about your claims—well, maybe not as shady as you want, but you can certainly be a late-night infomercial level of shady. Nobody really cares either, because the tech world exists more or less in a free market and if your product is garbage, 100 angry buyers have already written you a scathing Amazon review and anyone else can purchase your "half-star" product at their own risk. In many ways, it is a self-regulating ecosystem.

By contrast, the healthcare ecosystem is built on regulation. No one is going near your product unless you have strong scientific evidence to back up your claims. In fact, just ask the biotechnology fallen-giant Theranos how seriously government regulatory bodies such as the Food and Drug Administration (FDA) or Centers for Medicare and Medicaid Services (CMS) take fake claims. If you don't know about Theranos, Google it for an object (or abject) lesson in what happens when you don't carefully quantify your HealthTech product's claim, and not taking investment dollars based on fake claims and shaky science. If you're reading this and thinking "going the regulatory route is too hard, our product is a harmless <insert tech product here>, so we're just going to ask users to sign a waiver and sell direct to the patient" then let me invite you to read up on Lumosity. Lumosity is a company that develops "brain training" games in simple, sexy app form. The company was founded by good neuroscientists, allowing them to proudly use the tagline "backed by science" on all of their advertisements. Their product is simple and harmless enough—just download the app and play games. Although they had a lot of sub-scribed users, I guess they weren't making enough money and wanted to move more aggressively into the clinical market. So, they used marketing strategies to suggest that they might be able to help people with dementia despite the fact that MANY scientists were refuting these claims. It is crucial to note that these "marketing strategies" were not overt. In fact, they were incredibly subtle—for instance, if you searched "dementia" in google, an ad for Lumosity would pop up. No big deal, right? Wrong. Turned out to be a big mistake—the FDA needs to vet any company doing business in the US that claims to treat or prevent health conditions. They decided that even something as subtle using search engine optimization to place a Lumosity advertisement when you used certain disease-related search terms was akin to making these claims. Since Lumosity hadn't gone through the correct regulatory process, the Federal Trade Commission brought the hurt in the form a $50 million dollar fine, which was eventually appealed down to $2 million because Lumosity's financial situation was so bad at this point.

Why am I telling you these horror stories? Because I'm hoping that it will help you realize that you NEED a scientist on your team. A good scientist will quantify the effect of your team's product. Put simply, their job is to prove stuff. Like your engineer, a good scientist will prove just precisely what you ask them to prove, so make sure that you ask for the right thing. This goes right back to what we were discussing in Chap. 2 about constructing your idea: what is the problem that you're looking to solve? Once you have decided this, work with your scientist to determine the best outcome measures to validly and reliably prove that your product does what you claim it does. Check with your regulatory expert that your trial will hold water in front of the regulatory powers that be, and check with your end-users and market-analysis crew that these outcomes are actually meaningful and valuable to the intended market. Once you have checked all of these boxes, sit back and allow your scientist to conduct an unbiased, clinical trial of your product.

Now, in a moment we are going to go into a fairly comprehensive overview of how to run a good clinical trial; there are truckloads of textbooks that go into the minutia (not particularly interesting ones, but they are out there), but we will be keeping things fairly high-level. However, before we do, I want to take this opportunity to highlight a couple of common mistakes that I see many, many startups make. The most common move is when the founder of a HealthTech startup will swagger into a scientist's office and say, "Here's our amazing device, valued at $x thousand, and it's your lucky day! We're just going to just hand it over to you for free! The catch? No catch at all, just share your data with us." The HealthTech founder is charming and professional. The scientist is gracious and excited at the world of possibilities that the freely-offered innovative technology could bring. It's a match made in heaven. Then everyone parts ways, and one of two things will inevitably happen:

1. The device proceeds to sit in the scientist's office for the rest of eternity, and they eventually stop replying to the founder's enthusiastic emails inquiring as to whether they have had time to use their precious baby of a product on any patients (answer: no, they haven't).
2. The scientist excitedly uses your device as often and enthusiastically as they can and they send the founder TONS of data that was collected in an uncontrolled, frenetic manner. This seems great at first, until the company learns, after arduous and expensive consultation with the FDA, that they cannot use messy data to scientifically validate anything at all.

The outcome that *never* happens is that the scientist who is handed the 'free' product conducts a carefully considered, well-designed clinical trial on the founder's device. It's not that scientists are evil or lazy, it's just that getting evidence, real evidence, that a product works as advertised is really bloody hard, time consuming, and expensive to accomplish. You get what you pay for. I'm often asked, "why are clinical trials so expensive?" Well, to get a sense for why, let's break this process down step-by-step...

Why don't we bring it all back to our wonderful product (suite of products, really) that we've been developing in this book to help people recover from Total Knee Replacement faster. Let me set the scene: our fantastic HealthTech product team has been living near an orthopedic surgical facility that specializes in joint replacement for a while, and after doing all of the appropriate and responsible Research and Development they have made a mobile application that assists with physical therapy post-TKR. It is beautiful, secure when it handles health information (to the standard of the country you are in), multi-lingual (because you are wonderful, multi-cultural designers), and it does all the good functional things that it should, including:

- Tracks key patient metrics for the physical therapist.
- Gamifies therapeutic exercises to boost compliance.
- Gives real-time feedback to the user during exercise so that they don't push themselves too hard.

You have an exciting product, and you're eager to put it through its paces. HOWEVER, before you ever go near a patient, the first thing you need to prove is safety/feasibility. This means that some initial basic user testing has to occur to show that your HealthTech product is ready to be responsibly trialed on a protected population in a clinical setting. For each style of device, there are different types of regulatory hurdles. For instance, an app *may* have less potential to cause harm than a rehabilitation robot, but there will be basic standards to which you must adhere for both devices. This is where your regulatory expert is going to be crucial, so stay tuned.

Once you've proven that no one will be electrocuted, deceived, hacked or otherwise terribly maimed by your product, your next step is to prove the thing actually works in a clinical trial. Now, clinical trials are expensive and there are many outcomes that you can choose for your product. Remember what I said about scientists being like engineers? They will prove or disprove *exactly* what you ask them to–no more, no less. Work with your product team–your TKR experts–to decide on the *hypothesis* (i.e. question that you want to answer) of your study. Make certain that your question is one that, if proven, makes your product as valuable as possible. In the case of TKR recovery, at a high level there are a bunch of clinical tests that patients need to pass post-TKR before they're discharged. It's potentially of high value to an insurer (and a hospital) if you prove that using your app gets patients ready for discharge faster than with conventional care alone. Great, so this can be our research question:

> Does our app significantly reduce the average length of a hospital stay following total knee replacement surgery?

You typically only have one shot at a clinical trial, and it is hard to prove an outcome as broad as "patients who use our app simply get better faster," because recovery is so multi-factorial. A good clinical trialist will encourage you to pad your study with secondary outcomes that might also prove your value, without introducing any sort of unfair bias to the study. For instance, in this case some appropriate secondary questions to answer about the product could be:

- *Does the app significantly decrease the amount of time to independent transferring and walking?* Do patients using your app go from needing a nurse's help to move from lying in bed to standing, to not needing a nurse faster than people who don't use your app? Are they safe walking independently any faster? If your study shows that the app cuts down on the burden to nursing staff on a ward, you may well have something valuable.
- *Does the app result in better knee range of motion at discharge:* Maybe time to discharge isn't affected at all by your app, but at the time of discharge, people who use your app can move their knee more freely than those who didn't. This is meaningful because it indicates that people who use your app may be safer and more functional in the home environment post-discharge. A long-term follow-up study (more money) may even show that your product decreases readmission and complication rates (less problems).

OK. We're making progress–we have our product. It is safe. We know what questions we want to answer. Now, we have to turn to our scientist, who is hopefully not just a great scientist, but also an expert in the field, and work out exactly how we will measure our progress. What clinical tests, which have been rigorously validated by the clinical and scientific community, will we use to answer our carefully curated scientific questions? Choosing established gold standards is crucial, these metrics are usually far from perfect, but they are widely used and trusted. A lot of HealthTech products develop their own outcome measures to prove their effect. You can do this, but it is risky, and if you don't capture traditional, "gold-standard" outcome measures alongside your custom metric, people will be less willing to accept your findings—even if your clinically-unproven measure is more sensitive, valid and reliable than the traditional measures. It may seem unfair, but this is really all about cultural sensitivity. You need to learn to speak the

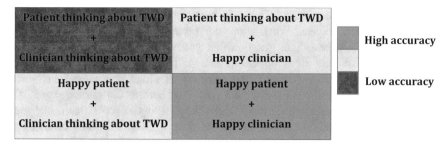

Fig. 4.1 Traumatized patients and clinicians influencing study metrics

language of the country that you're in before you try to change the culture…or else people will instinctively resist your efforts.

The other reason it's important to know field-specific outcome measures is because our next step in this process is to appropriately *power* our study. Regardless of whether you're asking the right question or choosing the right outcome measures, if you don't recruit enough patients, you can't make a strong case that your effect is not due to chance: even if the raw numbers of your study show a statistically significant difference. This is because clinical outcome measures typically work by having a clinician asking a patient to perform a task, watching the task, and then scoring the task on a scale of 1-to-whatever. This introduces variance into the clinical metrics, for instance, maybe one clinician isn't paying full attention during the assessment because they're still traumatized from last night's episode of The Walking Dead (TWD), or maybe the patient is feeling tired today (they couldn't sleep last night because of TWD). Then, maybe you have a patient and clinician who don't watch TWD at all, so they can focus on the task at hand very well. Then, of course, you can have any combination in between (see Fig. 4.1), which really does introduce a whole lot of noise to the system. Bottom line: TWD is horrific, and clinical outcome measures suck, so we need to spend time studying the *measures* so that we know how much they can be trusted. Good clinical outcome measures have already withstood a battery of validation studies. Thus, if your scientist knows the literature well, they will select outcome measures that have a track record for being reliable (many clinicians can run the same test on the same patient and return a similar score), valid (your chosen outcome measure is appropriate for measuring the variable that you want to track: for instance, you wouldn't use a patient's pant size to measure temperature, you would use a thermometer) sensitive (can detect even the smallest change in clinical status), and widely used (so that clinicians who are working in the trenches actually care about the changes you're showing). In the rehabilitation world, there is a great web resource called "Rehab Measures" (www. rehabmeasures.org) that gives a strong rundown of the most widely used outcome measures for tracking the symptoms of many different conditions, and how much you can trust each one. It is an excellent example of the level of understanding that you should aim for before you choose an outcome measure for your trials. Taking these factors into account, as well a few other features related to the type of statistical testing that you will be using, a good scientist (often in collaboration with a good statistician) will be able to tell you how many subjects you need to recruit to be confident of your effect on your desired population.

Brilliant, now we're ready to get started, right? Wrong. Now we need to discuss inclusion, exclusion, withdrawal and matching criteria. If you're comparing two groups (people using your app compared with people receiving conventional care), you need to think about all of the things that might influence your ability to answer your study questions. For instance, should we allow people with serious cognitive impairments into the study? Or people with diabetes severe enough to slow down the healing process? What about obesity? Or lung disease? Some of these factors can be controlled if you try to match out your populations so they have equal numbers of people with similar demographics and medical histories.

However, some factors have the potential to trash your study and mean that you need to carefully select your study population. You should also consider withdrawal criteria: what happens if someone enrolled in the study catches pneumonia and discharge is delayed by a month? Is it fair for that to mess up your data? Or should you write in a provision that if someone acquires an infection or has some sort of catastrophic incident that is not related to the use of your product, they are withdrawn from the study? A note of caution here: it is always tempting to get super-specific with a study target population in order to give you the best chance of success. The problem with that approach is that the task of recruiting *"60 white males between the ages of 55 and 65 who have just had a TKR performed but are otherwise in perfect health"* might end up taking you a thousand years, and it also means that you must limit your study conclusions to this group as well. So, be mindful of these issues, but strive for balance and acceptable risk.

We have now decided what we want to test, how we're going to measure, and who we're going to study. The next step is to decide on a study design. There are many different ways of conducting a clinical trial, and I'm not going to go through all of them, but I do want to give you a flavor for all the things you should consider. So, let's discuss one specific type, and really reach for the stars with what is currently regarded as the "gold-standard" study design for clinical research, which is a double-blind, randomized controlled trial (RCT).[1] It is not always possible to conduct this type of a study, depending on what you're trying to test. Wherever appropriate, if you don't go for the gold-standard, you will invariably be asked why, and if your answer isn't convincing you will be told "come back when you've completed an RCT". This is why having a good scientist on your team to navigate study design issues is so crucial; imagine going through the time and expense of

---

[1]Although an RCT is considered a gold-standard in many clinical research circles, experts are continually re-evaluating the best ways to quantify the effect of an intervention or technology. For instance, as the concept of "individualized medicine" is gaining traction, many fields have shown increasing interest in a style of experiment known as an "n-of-1 clinical trial". This form of clinical trial is very different to an RCT. Thus, it is crucial that the scientist on your team is familiar with conventional thought and research styles within the field in order for your clinical trial to gain the most traction.

running a clinical trial on your product, only to be told by a client/regulatory body that it was the wrong type of trial and your product efficacy evidence is weak. This is potentially a business-ending disaster.

Let's go ahead and break down exactly what a double-blind, randomized, controlled trial actually is. "Double-blind" means that anyone who is producing (patient) or collecting (clinical assessor) data in the study does not know if the research participant is in the "intervention group" or the "control group" (separate clinicians handle the patient's therapy in order to keep the assessor blinded).

Blinding helps you control for biases that may emerge when people get overexcited about being part of a clinical trial–patients can get better more easily (placebo effect) and clinicians can even score patients differently based on their own internal biases related to whether they think the trial is going to work or not. "Randomized" means the researcher doesn't get to choose which patients go in the intervention or control group. Patients are recruited and then randomly assigned to one group or the other. This prevents researchers from accidentally or intentionally putting patients with the best chance of success into the intervention group. "Controlled" means that your trial has a "control" group and wherever possible, you want to compare your intervention to standard practice. The group that is receiving standard practice is called the control group. This allows us to measure not just whether people using the new intervention get better, but also to prove how much better (or not) the new intervention is compared with conventional care, which is really the most valuable thing that you can prove about a HealthTech product. As a final point on all of this study design stuff, if you're blinding your patients to whether or not they're receiving your app as an intervention, you need to get a little creative with your control group. Either you don't tell your research participants what the trial entails (meaning your clinical trial now involves a deception, which sounds more sinister than it is, but is still taken quite seriously by the hospital ethics committee), or you need to design a whole other app that looks a lot like your app, but doesn't provide any clinical value (this is the harder, but more scientifically rigorous option).

In our TKR rehabilitation case, we could certainly conduct a double-blind randomized, controlled trial. Here's how it might play out:

1. You recruit and solicit informed consent from patients who are coming to your hospital of choice to receive a TKR.
2. In order to keep patients blinded, you must choose option 1 or 2 listed above. If you choose option 1, you must remain fairly vague about the specific intervention, while being incredibly transparent about all of the potential risks and/or benefits that being involved in the trial would entail.
3. Post-surgery, recovering patients who have consented to be a part of the study (and meet the inclusion/exclusion criteria) are randomly assigned to either the control or intervention group.
4. A blinded assessor will collect any baseline metrics relevant to the trial (for example, knee range of motion and some functional measures).

5. The trial begins, our control subjects receive conventional care, our intervention subjects receive conventional care + our amazing TKR rehabilitation app.
6. The blinded assessor collects data at however many points in the study your scientist thinks is appropriate. In our case, this would likely include collecting a set of metrics at discharge, and probably even a 6- or 8-week follow-up (to look at rates of readmission, falls, longer-term outcome, etc).
7. When we have met our recruitment goal for the clinical trial, our scientists crunch the data, compare outcomes between the control and intervention groups, and hopefully show that people who used the app have a statistically significant improvement in outcomes compared to people who don't.

Ok. Even I'm getting impatient now—can we get started, already? Why do we have to drag ourselves through all these details before we've even looked at a single patient? Because, your entire research plan needs to pass muster with the ethics review committee of the hospital or university where you plan to do your research. To my American readers, you may hear the term (usually spoken with great dread) "Institutional Review Board" or "IRB" to describe the group of individuals that are about to hold the life or death of your product in the palm of their hand. You need to convince a room full of people that not only is your device safe, but that the experiment you wish to conduct has value to the field, and that value outweighs the potential risk to the patients. There is no way to conduct legal, ethical research that can be used to validate your product to any regulatory body anywhere in the world without first receiving approval from the institutional ethics committee attached to your study location. Not gonna lie to you on this one, the process is pretty brutal. Prepare to drown in paperwork that interrogates your study from every angle to ensure that patient safety, privacy, and dignity are preserved through the course of your study.

People on these committees take their role extremely seriously (as well they should), and they will absolutely rake you over the coals if they think you're trying to do something underhanded or unsafe. Be aware that this is a lengthy process, so start planning with a lot of lead time. In most hospitals I have worked in, these committees meet every month, but because they meet so infrequently, they are usually backed up with cases. Thus, prepare yourself for the possibility that you may be on a waiting list that is several months long. Make sure your scientist fills out your paperwork as carefully and precisely as possible because if there is even the slightest mistake, the members of the committee will ask you to correct it, and then they need to re-review it before approval. This whole process can get pretty tiring, and can delay your clinical trial for months if you don't plan it out carefully, which can absolutely kill a startup at a vulnerable period. The shortest it has ever taken me to receive ethics committee approval for a study was 1 month. The longest process I've ever been involved in was 9 months. I want to point out that for the application that took 9 months, it didn't take long because I was proposing a particularly risky trial, or trying to be sketchy or unethical. Nope, it was all because the institute I was working with had some deeply incompetent bureaucratic practices surrounding the whole ethics approval process, so it took us forever to get permission to start our project. The good news is that I have had these experiences so that you don't have to! If you're targeting an institution, make sure your due diligence includes asking researchers about the length of the ethical approval process. It could end up saving you a significant amount of time and money.

I hope that at this point, I have really impressed on you how important it is to validate your product in a scientifically rigorous manner. I also hope too that you have started to gain a heightened appreciation of how time is money for a startup, and avoidable delays can be devastating. As a HealthTech startup, you have to pick your pilot institution very carefully. If I had to rank it, I would say that the three things that matter most when you're selecting the institution that will conduct your research are:

1. The ability of the institution to meet your patient recruitment needs.
2. The institution's ability and readiness to conduct scientific research (this includes the length of the ethics review process, but also general level of experience and interest of the research staff for conducting research).
3. The reputation of the institution (good is better than none, none is better than bad).

I know that this section has been long, but it is crucial to deeply understand the role of science in HealthTech. This is what makes building a HealthTech product different from building the next Candy Crush. All of the money projections about the massive size of the HealthTech market are real, but this is the catch—I'M EXPLAINING THE CATCH—you **must** rigorously prove your product is capable of helping people. It will take time, it will be expensive, but you can't create a successful HealthTech product without this level of assurance.

Regulatory Expert

Many HealthTech companies that make it through the challenging gauntlet of product design and development will still end up failing because they did not attend to the need for regulatory approval until it was too late. Regulatory approval can introduce incredible costs and delays to your process. You need to budget for these costs when you're finding funding for your product, or you end up with a product that is not allowed to be used on patients and a bunch of angry investors wondering where their money went.

In our last section, we covered the process of scientifically and clinically validating your product. I really wanted to impress upon you how detailed and involved that process can be. The reason you need to complete the science is so that you can approach regulatory bodies for approval in a way that is evidence-based and systematic. The regulatory processes surrounding healthcare in most countries is nothing short of brutal—endlessly complicated and very particular. Thus, someone on your team who is an expert in completing regulatory procedure is a must-have. This is the person who is going to tell you what will and will not fly with the regulatory body that you're working with. You want your regulatory expert involved in every step of the process, with every member of your team and as **early** as possible. So, make sure they speak to your:

- **Engineer**: *For a device*—what components will you be using? Are they appropriate for a "medical-grade" product? *For software*—what cybersecurity measures are you taking? How are you designing your product for ultimate system stability? What are your contingency plans/what are the consequences if your software crashes?
- **Designer**: *For a device*—what materials will you be using? Are there infection-control measures required? Can it easily break and create a sharp edge? *For a software*—what sort of language/content are you using? Could it potentially trigger someone?
- **End-User**: Under what conditions will you see yourself using this product? Is it going to be in public? In private? Only in a clinical setting? What particular use scenarios could feasibly introduce risk into the process?
- **Scientist**: What do we need to prove, and how well? What effect sizes are we expecting in the clinical trial, and will they be enough? What is the burden of proof in regulatory circles for particular product types? How serious are the different types of potential adverse reactions, and how do we handle reporting them?

As you can see, your ideal candidate for this role will be a highly engaged and heavily involved regulatory expert to keep an eye on your product in order to save you heartbreak in the long run. It should come as no surprise to you that many HealthTech companies really don't do this well, because they fall back on the tech product mentality, i.e. "this is just another tech product, and we're going to treat it like one." Please don't fall into this trap, because it won't end well for your company. Your product cannot help anyone until it has been approved by a regulatory agency. This is the part of your product that will probably cause you the most sleepless nights, but if you achieve regulatory success, it will immediately differentiate you from many competitive products in the marketplace, because it is currently the aspect of the HealthTech industry that is most often overlooked by startups.

I hope that this chapter has given you some valuable insight into the anatomy of a HealthTech team. As with all things anatomical, your team is going to be special and unique in its own way as you embark on your HealthTech venture, but what we have covered here is intended to give you a solid groundwork. In the last four chapters, we have covered the essentials of HealthTech product formation. In the next section of the book, we're going to cover some specific case studies in greater detail.

Part II
Case Studies

"Success is stumbling from failure to failure with no loss of enthusiasm" —Winston S. Churchill

If you've made it through the first part of this book, and are *extremely* perceptive, you may have noticed that I have a lot of opinions. With that in mind, this section of the book is to give you a small, but sufficiently dangerous, insight into my psyche—mainly, some of the experiences and projects in this emerging space that have shaped my perceptions of the HealthTech ecosystem. I've tried to give you a real taste of products from varying perspectives, so in this part, we're going to dive into case studies of some of my favorite teachable moments in HealthTech involving large corporations, tiny start-ups, elite athletes, government initiatives, and social good enterprises. I've made a point to select a few stories of failure, because we really need to dispel any sense of Survivorship Bias in the HealthTech world. In emerging fields, it is very easy to fall into the trap of making up the rules as you go along, while having a tendency to study only the success stories in your field, since you rarely have visibility into failed companies. The grim reality of the HealthTech world right now is that far more companies will fail than succeed. Thus, there is tremendous value in studying projects that did not work out so well: understanding what the pain-points were, and why and how the project failed. In some of these cases, I will be changing or omitting names in order to protect the innocent (or the guilty!). I also want to be clear that even in the cases of comprehensive failure, I'm still quite close with many of the team members in the stories because they are good people. We just got caught up in a product that didn't work out. In many ways, I think that a great factor in determining whether you want to keep working with a person or not should be how they handle failure. Failure is common in emerging fields, but acting like a spoiled child that dropped their ice cream when your project doesn't work out, is not. Astoundingly, though, I'm happy to report that not all of the stories I relate here will be harrowing tales of failure. There are a couple of happy endings in here. What I can promise is that all of the case studies I present will have some form of actionable takeaway that relates something valuable about the HealthTech industry.

Chapter 5
The Big Corporation

"If you can't feed a team with two pizzas, it's too large."
—Jeff Bezos.

The first case study that I would like to share happened in my first year of working in my position at Burke Medical Research Institute. I had just accepted the position, and within about 3 weeks I was thrilled to find myself sitting with a high-level executive running the North American arm of a massive, "household name" electronics company. Now, I know that to some, a scientist teaming up with a big business might come off as a little unseemly. So, before we go too much further, I want to point out to anyone who may not have heard: being a young faculty member at an Ivy League institution in the US is really hard and stressful. In this context, you are essentially handed the reins to a startup where you're trying to sell your product, which is your highly specific brand of science. You are told by the university that they will give you everything you need to conduct your research for a couple of years, but by the end of a 2–4 year period (depending on what you can negotiate), they expect your lab to be entirely self-sufficient. What this means on a pragmatic level, is that they expect you to pay for the salaries of all of your employees and yourself, pay for all your equipment, pay rent on the space and facilities you're using, etc. Except, here's a twist: the biggest funding body in the land (the National Institutes of Health) is funding all of your competitors and is also so broke that they are unable to fund anything but a tiny percentage of research projects (which tend to be selected, near as I can surmise, not on merit, but by painting your grant number on the back of a rat and racing your rat against several thousand other rats in the bowels of a government building in DC. I believe there's a documentary on Netflix…). Sorry to take such a long aside, but the point is that it is tough out there, and science needs all the help it can get. Where were we? Oh yes, I was incredibly happy to be sitting down with a powerful executive that could potentially fund a lot of my research.

This person explained to me that their company had recently taken a keen interest in telemedicine. They were acutely aware of the fact that populations in developed countries around the world were aging at an unprecedented rate, and their parent company had investments in many aged care facilities and hospitals,

© Springer International Publishing AG 2018
D. Putrino, *Hacking Health*, https://doi.org/10.1007/978-3-319-71619-0_5

and they saw some really great synergies to be had if they started to develop products for "successful aging." They had done some detailed market research, which had cost them a lot of time and money, and they had made the decision that there was a lot of opportunity in the Alzheimer's market. Alzheimer's Disease (AD) is a disease of aging, which means it is on the rise. At the time of writing this, the Center for Disease Control (CDC) estimates that the costs associated with caring for seniors with AD in the US alone is 259 billion dollars per year. The Alzheimer's Association projects that by 2050, AD will cost the US more than 1 trillion dollars annually. With these factors in mind, the company developed a mission: use technology to make life easier for people with AD and their caregivers. The notion is quite solid from a business perspective: this is a costly disease, so if you can use inexpensive technology to decrease caregiving costs by even a small percentage, you will have a product with enormous value.

The company worked on a product to achieve this goal, and the end result was a tablet with some very specialized software loaded onto it. Harnessing the staggering resources of a company their size, they hired FDA consultants and HIPAA security experts, and created a product that was annoyingly perfect from a regulatory standpoint. I also really loved their user interface and user experience. The front end was neat, clean, and designed with seniors in mind. The back-end (where people entered information for the person with AD) was slightly less intuitive, but was still no more complicated than entering bio info into Facebook or LinkedIn. For cyber-security and logistical reasons (that made the most sense to the company), they decided that the product would be both a hardware and software hybrid rather than just a software app. The goal of this product was to create a social interface and organizer for the person with AD. They could input social events, medication reminders, appointments, etc. However, approved individuals with login credentials (such as immediate family members) could share photos, videos, and texts through the interface. The real "dream" for this technology was to have every senior diagnosed with early AD of Mild Cognitive Impairment to be assigned one of these tablets. If it worked as advertised, it was hoped that this device would make them more organized, improve their quality of life, decrease burden on their caregiver, and maybe even help them to feel less socially isolated.

So, the company made their first foray into testing out the technology. They provided the tech to a care facility that served many clients with AD. They provided the tech to the facility, helped the facility staff enter in information for each user, and then a whole bunch of seniors had access to this technology. And they loved it! They seemed to enjoy using it every day, receiving photos and videos, and they found the user interface intuitive. They didn't want to give the devices back at the end of the trial! It led to an unfortunate mass-tasing incident at the test facility (no, it didn't). But, all the things that you hope to hear in an initial trial, the company heard. They came to me, very confident and excited about running a more detailed trial for their target population. However, now they were very interested in focusing on what they considered to be a larger market: community-dwelling seniors with AD or mild cognitive impairment.

One of the most important things I learned during this project was how differently people coming from the business world think about issues compared with people in the clinical and scientific world. Before we go too much further, I don't want you to get the wrong idea about me: I don't take any sort of delight in ruining anybody's day, and I'm not falling over myself to tell people their products suck, but sometimes a dose of reality is important. This company's first pilot trial was cool, but was it really a good indicator that their product was going to actually work across the broad population they had in mind? For instance, was it reasonable to assume that for each unit they sold, someone from the company would head over to help with the setup (like they had in the care facility)? Is an enjoyable product something people with AD will pay for? I mean, forgive me for asking, but do we even care if the users are enjoying the product? Are there other outcome measures that we should be tracking that are more important to selling units? This barrage of annoying product questions is exactly why I am no fun at parties. But if a large corporation has the means and willingness to save the world then I'm certainly going to do everything in my power to help them. So we decided to conduct a clinical trial of their technology together.

The corporate-level management team at this company were really good people. We sat down, and they listened–I mean, **really** listened–to my scientific ramblings. We discussed conducting a clinical trial at Burke, where I was working at the time, because Burke happens to have a first-rate clinic for AD called the "Memory Evaluation and Treatment Services." The facility is unique for all of the services they provide, and the bleeding-edge research that they conduct. I had a chat with the wonderful team at the Burke clinic and they agreed to help me run the trial. We sat down with the company and asked them what they wanted to prove. Like every technology company entering a medical space, their answer was pretty modest: they wanted to show that their product completely transforms the lives of people with AD. That's all. Should be no problem, right? It sounds funny, but I often feel as though the thing that makes me good at my job is my ability to recalibrate the expectations of a highly enthusiastic product team without completely destroying their souls/having them lose faith in themselves and humanity in the process. We sat down and tried to operationalize our thinking around the product by thinking of it in terms of some clinical/technical outcomes that can be measured:

- Quality of life for the person with AD.
- Quality of life for the primary caregiver.
- Caregiver burden.
- Social isolation for the person with AD.
- Product usability metrics for both the caregiver and person with AD.

Great. At this point, I've got to say that we're looking at the holy grail of HealthTech product development: a large corporation had put significant care and thought into developing a product, and they spared no expense in ensuring it would pass regulatory muster if we just prove that it has some form of valuable effect on our target population. We identified the things that we wanted to measure, we had convinced a stellar clinical partner to play ball, and we even found a small budget to support the trial. Next, we developed a protocol we thought would be pretty straight forward:

1. Intake: Recruit prospective subjects from the clinic at Burke, screen them for eligibility in the trial (like all good clinical trialists, we had an extensive list of inclusion and exclusion criteria for the trial), and for eligible, willing subjects we would obtain informed consent, and collect the baseline measures for all the outcomes previously identified as important.
2. Eligible subjects are then introduced to the technology, shown how to use it (with very standardized, practiced instructions) and told to take the tablet home for a month to use as much (or as little) as they liked, and then bring it back to us at the end of the month. At the end of the month they get a small gift voucher ($50) for their trouble.
3. At the end of the trial, we bring together all the metrics we collected at baseline, and decide if the product had added measurable and tangible value to the lives of people with AD.

We ran this protocol by Burke's ethics committee, and all-in-all they concluded that it was a pretty low risk, while still clinically interesting, trial. They gave us permission to proceed. On our end, we formed a team to carry out the trial, and the corporation assigned an employee (who was part of the engineering team for the product) to coordinate things on the tech side, i.e. troubleshoot any issues that we or our research participants may be having, fix bugs, the usual stuff. We also planned to join in a standing weekly meeting to discuss anything that might come up. I'm not the biggest fan of standing meetings, but I am a fan of open communication, and funded research, so, alright, we'll have the meetings.

We had carefully prepared, dotted all the 'i's, crossed all the 't's and were finally ready to go. SPOILER: this is where the unmitigated disaster begins. It all started off innocently enough with our team canvassing the clinic to recruit subjects for the trial. Upon learning about the trial, most prospective participants were immediately put off by the idea of being given a piece of hardware, even for FREE! I don't run an electronics brand, so maybe I don't know the business, but in my non-business mind, this started to set off a few alarm bells. Meanwhile, Richard (totally not his real name), our friend from the company, was excitedly emailing us about our first weekly meeting. The call went a little something like this.

Week 1:
Richard: Hey guys, how's it going?
Us: Not great. We've probably approached 50 people so far, no one really wants to do the trial. They don't like that it involves a new piece of hardware.
Richard: This was always going to be a challenge—seniors aren't that comfortable with technology (**sorry to break character, but this is a false statement that is surprisingly widespread**). *Really try to work on your salesmanship with them. They'll try it out if you sell it to them right.*

Us: Yeeeeeaaahhh, not sure that this is really the case, Richard. Our senior community gets technology, they've just looked at this technology and they don't understand why it can't just be an app they download on their smartphones. They're even asking if they can get it in iOS, because they don't like Android. The people that we've approached seem pretty sophisticated with their tech knowledge. They really want an app.
Richard: We've discussed this in our focus groups. That is not what seniors want. Seniors want a device.
*Us: Oh, OK. Good. Thanks for setting us straight on that. See you next week! *click**

Not the best first week that one could hope for, but, you know, they *are* a big company, and they *did* use focus groups. Maybe our first fifty interactions were a bit of a fluke. All we can do is keep trying. So, we kept approaching people and recruiting our little hearts out. After a few more weeks, I didn't really think the first 50 were a fluke, I think they were pretty representative of how everyone felt about this thing. That was fine, though. We felt we could still continue the trial, evaluate the product, and if we saw an exciting benefit from using the system, the company could always "appify" after the fact to deal with the main critique that the seniors had about not wanting to commit to another device.

The weekly meetings didn't get much easier with less than impressive recruitment numbers trickling in, but we did what we could, and we slowly started to get some people to agree to test the product. After we had recruited a few people we started to notice a workflow issue arising, which we discussed with Richard:

Week ~10:
Us: Hi Richard, wanted to run something by you.
Richard: Shoot.
Us: Well, it's about our clinicians—they had some feedback about the product. As you know, they're happy to help us test the product in the context of a clinical trial, but long-term, if this were to be a product or service in their clinic, they're concerned about who exactly would teach the patients how to use the tablets.
Richard: The clinicians would! They can offer it as a service, and if the patients want the service, they can teach them how to use the tablet
Us: Right. Yes, but it takes about 20–30 min to walk a patient and their caregiver through the system—our clinicians are pretty busy, and they are concerned that they won't be able to spend that amount of time on each patient. Outside of a funded clinical trial, who will pay for their time?
Richard: Our product testing shows that a <u>competent</u> clinician can teach a patient the system in 7 min.
Us: Weeeeelllllll, firstly, our clinicians are pretty awesome, and we totally agree that the content doesn't take so long to explain, maybe 10 minutes-
*Dick: **7 min***
Us: But that is the time it takes once the patient is sitting down in front of the system. In terms of clinical workflow, getting the patient and caregiver together in a

room, getting the tablet in front of them, dealing with questions in a way that creates an experience where our patients don't feel rushed and uncomfortable—we've been timing the full interaction, and it clocks between 20–30 min.
Dick: Once you get used to our product you'll create more efficient clinical protocols to accommodate it.
*Us: Oh, OK. Good. Thanks for setting us straight on that. See you next week! *click**
Also Us (once the phone was hung up): Wow.

The trial continued on, and things really didn't get much better, but the recruits trickled in and we collected data piece-by-piece. We tried all the tricks we could think of: fliers, hiring an additional clinical coordinator, getting the local Alzheimer's Association charter to advertise the trial, and we attended local AD support groups to pitch the trial. Dick (*still* not his real name) grew increasingly impatient and perturbed by our supposed lack of progress, even to the point where he called one of my interns on his cellphone outside of our scheduled weekly meetings to interrogate him as to whether the team was trying to tank the trial for some reason. Dick was sure that this was why our recruitment numbers were so poor. At this stage, I want to stop and point out that I'm not saying all of these things to be disparaging: I want this to be instructive, and there is nothing quite so soul-destroying and existential-crisis-inducing as a failing product trial (there's probably a few things, actually, but the point is that it is incredibly painful to watch a product you care about fail in clinical trial). However, there are good ways and bad ways to deal with crisis. If you remember to be mindful in the crisis, you can reframe the experience and pose questions like "this is not going so well, but what can I take away from this experience?" OR "How can I pivot to make this disaster a new opportunity?" If you're not so good in a crisis, the response tends to be to remain myopic, like our buddy Dick, and the tendency becomes one of assigning blame.

I had a lot of young members from my lab working on this team, and they were distressed that the trial was doing badly. They were especially upset that the representative of a big, powerful company seemed pissed at them. All of this is an extremely unfortunate side-effect of a trial that is not going well, and for my part, I worked to continually remind them of our lab's mission: we're interested in studying the interaction of digital health technology with the clinical world. From our perspective, a failure is just as informative as a success, so put Dick out of your mind, and continue to collect data on everything that is going wrong. This is a unique opportunity. So, we persisted. We collected data, we listened, we learned, we documented. I'll share one last phone call that sticks out in my mind, because this one was so crazy that I remember at the time it felt almost surreal:

Week??20?? (we are over-time, over-budget, no-one is happy)
Us: Hi D-Richard, how's things? We wanted to let you know that we've been running into an issue on the caregiver back-end
Dick: Ok—what is it.

Us: (I'm going to keep this vague because it was a super-specific issue) *Well, all of our users are getting confused about how to fill in data on this tab: they think it is asking for information piece "A," but really it's asking for information piece "B." They keep getting confused, so we're thinking you could change it in this way to alleviate confusion. Our research participants agree that this solution would make the workflow clearer to them.*
Dick: We discussed this issue in our focus groups during product development. It's not a concern.
Us: Uh, but our users are telling us it is a concern. They are *specifically* citing this issue as a reason for marking the system down on its usability.
Dick: It is not a concern.
Us: Oh, OK. Good. Thanks for setting us straight on that. Speak to you next week! *click*

OK, I don't think you need to hear any more (loosely) transcripted phone calls between Dick and my team—You're smart, and I think you see what is happening here. Dick forgot to listen to the end user. We were giving him absolute gold in terms of feedback, but he didn't want to hear it, so he instead cited focus group findings, or beta-product testing data as reasons to ignore our feedback. Now, with 20–20 hindsight it is very easy to sit in our armchairs and think "what was Dick thinking?" However, as someone who has been in Dick's position, I can tell you what he was thinking one of two things:

1. *"We've already done this to death and I don't want to do another redesign; the engineering team will kill me."*
2. *"If we change our platform every time a small trial gives us feedback, we'll never stop changing our platform, and who knows if it will even help"*

One of these perspectives is indefensible (thought #1), and one of them is actually fair, and a hard thing to gauge. However, if I'm going to guess, I think Dick was thinking point #1, and the product suffered badly as a result. A second thing to learn from this experience: the big corporation was not entirely to blame on this front. At the executive level, I did (and still do) really like and respect the people that initially brought the project to me, but assigned the wrong person to take the lead on their end, and gave him very little oversight. This turned out to be a mistake. I often wonder what could have become of that product if a more skillful and proactive lead had been assigned to us. As it stands, however, it should surprise nobody reading this that it is no longer a product. Several million dollars and 3–4 years of work later, the company finally sent it to their R&D graveyard of ideas that never really took off.

I know you probably want to look away at this point, but I'm not done with the post-mortem just yet. There is more failure to talk about (hooray!) and learning to be had. Let's dive into what our users thought of the product, because we had some really mind-blowing insights here that I think are generalizable to many products in this space. After our high-level interactions with the company, I think the next logical step is to quickly talk about the clinician's perspective on the product.

Our clinicians had no problem using or understanding the technology, and on a general conceptual level, they really did like the product. However, they did NOT like having to spend time explaining the product to patients. If your product needs 20–30 min per patient, which clinicians can't bill for, then your product is a non-starter. Clinicians are busy and overworked. Every second of their time is accounted for and billed for in most health networks (even socialized ones!). This is such a crucial issue, which I've seen emerge and destroy many promising HealthTech products. I really can't emphasize this point enough: *even if your product improves patient outcome*, *if it costs clinicians time per patient that they cannot reimburse*, *they will not use your product.* Often it won't even be the clinician's decision, but rather policy of the place where they work. This is a really shameful part of most healthcare systems, but it is a reality that we must confront when we are designing healthcare-facing products.

Now, let's talk about our research participants. First, let me give you a sense of the numbers we managed in the trial so you can get a sense for the initial and eventual scale of this doomed project. Over the course of about 7–8 months, we approached almost 200 seniors with mild cognitive impairment or AD. From this large number of people, only about 18 people ended up showing up to the clinic to complete the screening and baseline assessment for the trial. Of these 18 people, as we were explaining the technology in more detail, 10 dropped out before even beginning the trial. They didn't like the technology, and didn't want it in their home. I want to remind you all at this point that the protocol dictated that we hand them the tablet, free of charge, they hold it in their home for a month—no obligation to use it—and then they hand it back to us in return for fifty bucks. **Nope**. **Turns out**, **we could not pay people to take this tablet**. When we pressed the 10 people who dropped out for reasons why, the story was typically the same: the primary caregiver does enough and they don't want to have to learn another system. I know I'm not surprising anyone here, but AD really is such a hard, cruel disease. Your primary caregiver is typically your spouse or child, and they are watching something truly horrific happening to someone for whom they care deeply. Primary caregivers in AD are typically overworked, stressed, and going through an impossibly complicated grieving process. In addition to this, if they are the spouse of the person with AD, they're probably pretty old and have their own things going on! When you take all of this into account, and then send a perky intern in to train them on how to use a new, unfamiliar computer system that doesn't promise to cure AD or save them money, I guess we shouldn't have been all too surprised that most of them told us to hit the bricks. From the ten participants who dropped out, we heard the same sorts of sentiments:

I already do so much for [partner/friend/parent], I need to do this as well?

I write things down on paper or on my phone, why do I need to learn a new system?

How about designing something to support me? I already give [partner/friend/parent] everything that they need, I'm the one that needs help!

I found it really interesting that a product with such altruistic intent could bring out these reactions, but when you ruminate on them you start to realize that these are really valid and logical responses to the proposed tech. If your product is going to solve a patient's problem, but creates more work for the caregiver, you better be sure the problem is damned important to both of them.

Out of ~ 200 solicitations, we're down to 8 people. Well, subtract the two that failed screening. So now, after all this time, work and expense, we have 6 people enrolled in the trial. Clinical trials are hard, sometimes. It gets better, though. One of our participants hated the product SO MUCH that they withdrew from the trial 2 weeks early! They gave up their $50 participation gift just to get the product out of their house faster. Needless to say, my team (that loves a good failure) was *super* interested in this couple. We sat down with them to discuss. Now, recall that the company had insisted on creating a hardware product, and to that end, they created a custom tablet. I have to admit that I thought the design was beautiful: it was nicely colored, it looked kind of futuristic, it came with a little stand, and it was portable. It looked very unique and impressive, which, as it so happens turned out to be a problem. This particular couple was rather socially active and had people visiting their house quite often. They observed that the first thing to happen when people visited, is that they would point at the new, impressive-looking tech and say "Wow! What's that?" They would tell their visitors. Then, in their words, *"and all of a sudden we're talking about his Alzheimer's Disease rather than enjoying a nice visit."* The device, with all its beautiful, customized design, became a constant reminder in the house of our participant's disease state. He hated it, and it had to go.

Then there were 5. Our remaining five subjects made it through the trial without incident. None of them experienced significant benefit (or even some sort of a trend toward benefit) in any of the domains that we tested. None of them were particularly impressed by the tablet and would not have opted to purchase it if given the opportunity. The reviews were pretty lackluster, but my favorite statement in the post-study interview was from a thoughtful old gent who was a retired engineer (I should probably just say 'engineer,' because engineers never retire from thinking like engineers!). He said simply, *"Your product asked too much and gave too little."*

I can't be more cogent or eloquent than this amazing man. Read this message and take it to the bank with every product you ever develop. The clinical, social or monetary value your HealthTech product brings to the table needs to be greater than the effort required to use it.

Chapter 6
The Small Startup

"I have not failed, I've just found 10,000 ways that won't work"
—Thomas Edison.

If the large corporation failed, surely the best pathway to success would be a small, nimble startup, right? Just kidding! Of course not—this is another story of failure. But in fairness to me, it is failing in an entirely different way. So there. However, this story starts with a great pitch as well: a team of biomedical engineering students from the City College of New York had to complete a senior design project for their class. The students in this class are given a pretty clear directive: go out, find some clinical researchers and build them something that will help their patients—City College really has an awesome program in that regard, and I still collaborate with the school to this day. So, enter three young biomedical engineers who approached me and asked if I had any ideas. In this case, I was the one doing the pitching:

Stroke is the leading cause of permanent motor disability in adults. Stroke survivors can benefit from rehabilitation but often have limited access to rehabilitation due to logistical difficulties. If we could use emerging technologies to effectively improve access to rehabilitation in the home environment, we could reduce costs of care delivery and improve long-term outcome of stroke.

The idea was simple enough, and I had recently become enamored with a brand new piece of technology—the Leap Motion Controller. The Leap Motion was (and still is) a pretty revolutionary piece of hardware—it is this cute little box that costs under a $100 and is embedded with a bunch of infrared emitters and detectors. When you wave one or both of your hands over it, whatever black magic algorithms the developers have cast on this little device allow it to detect exactly what your hands are doing in real time. This device is pretty impressive, and if you're interested, you can check out some pretty cool videos of what it can do on YouTube. At the time, the Leap Motion ticked a lot of boxes for me. First, it was cheap—gotta love that. Second, the software was slick and hard to crash—an absolute must. Third, it came with a very generous Software Developer Kit (SDK) that allowed you to do whatever you want with the data streaming in from

© Springer International Publishing AG 2018
D. Putrino, *Hacking Health*, https://doi.org/10.1007/978-3-319-71619-0_6

the sensor—cool. Finally, it passed what I liked to call the "Amazon Prime test", which is my highly scientific metric for technological accessibility.

Let's launch into a quick side note about this, because I think that accessibility is a practical requirement of many HealthTech products that is often overlooked. I have seen a lot of good products fail because they couldn't scale fast enough after showing efficacy, either because their hardware was homegrown and their manufacturing pipeline was not set up effectively, or, they selected a hardware partner that couldn't deliver. If your HealthTech product is primarily software, but needs to interact with a device in order to work, then you have a hard choice to make: do you attach yourself to a particular piece of existing hardware, or do you build your own? Well, most HealthTech startups would agree that if your aim is to develop software, then don't spread yourself thin—let someone else design the hardware, and you piggyback onto their product. In addition, that means that you can move fast: your software can be in every home in the country overnight if it really takes off, hardware takes longer. Therefore, if it is important that your product can move at the "speed of software", make sure you select an accessible hardware partner. Understanding a hardware product's accessibility can be complicated—not all companies are incredibly transparent about the availability of their hardware. Which is why I developed my test: if I can hop on Amazon (or any other online shopping service), order the product, and have the option to receive it within two days (or same day!), then I'm looking at a product that has sufficient manufacturing and distribution capabilities to scale with me as my software scales. Obviously there are exceptions to this rule and you should always do more research, but as a general guideline it rarely lets me down. Speaking of general guidelines, on the flipside—be very careful about aligning yourself with any product where, when you click 'buy' on their website, you're directed to a phone number or email address rather than a shopping cart. This is usually a red flag. Finally, I wish I didn't have to tell you this, but I've seen it so much that I'm just going to say it in case any of you readers are thinking it: DON'T align yourself with a Kickstarter product…WHILE THEY ARE STILL ON KICKSTARTER!! There are few guarantees in this life, but this move is a disaster every single time. I don't care how cool it is: wait for them to become a product before you align your product with it. Sorry for that, perhaps unwanted, forced march down "accessibility" lane, but it is an important, often overlooked feature of your product strategy.

Now that is out of my system, let's get back to the story. In collaboration with these students, we developed a video game that was designed to rehabilitate stroke survivors. The video games were movement controlled: the Leap Motion Controller was able to identify and track movements that were relevant to rehabilitation of the wrist and hand, and the game created a fun environment that encouraged stroke survivors to complete many repetitions of these movements. We designed an interface that allowed each user to have their movement capabilities assessed prior to starting the games, that way we could customize the difficulty of the game to the severity of the stroke survivors' symptoms. Finally, we designed a therapist back-end that allowed them to track their patients' metrics related to compliance and movement quality over time. I was pretty satisfied with the prototype we had made, and it would seem that so were a bunch of tech entrepreneurs, because we won second prize in an entrepreneurship competition. This landed us $50,000, a year of free workspace in New York City, and a year of free legal and regulatory advice to help us incorporate a business (as first-timers, we had no idea what we were doing on our own!!) and get the product ready for market. Not a bad little prize package. The team members were so grateful for my help in the development process that they made me a co-founder of the company, which they rapidly incorporated as a company named 'GesTherapy' (short for "gesture-based physical therapy"). In addition to a co-founder title, my fellow co-founders gave me some equity in the company in return for my help in scientifically validating their product so that we could start to make some claims that potential customers would find interesting. This all turned out to be pretty great timing because I was just starting my position at Burke Medical Research Institute, giving us access to a large pop- ulation of stroke survivors who were interested in trying out new approaches to stroke rehabilitation. Over the course of the next 12 months, we ran a clinical trial with positive results, and conducted feasibility studies with both clinicians and stroke survivors showing that they found the technology usable, intuitive and enjoyable. We received press from some major media outlets internationally: in print, TV and radio, GesTherapy was being blasted in Australia, the US and the UK. Our phone and email accounts were blowing up with stroke survivors asking to purchase our product. On the academic side, we were getting some attention as well —we had some peer-reviewed publications under our belt, we were invited to present data at scientific meetings and I even presented some of our data as part of an invited presentation at the World Congress of Neurorehabilitation in 2016. All signs were pointing to the idea that we had a good product and the momentum to maybe actually make it as a company.

This is where things start to get a little sad, though. As I have mentioned before, clinical trials are hard and can be incredibly expensive. We had completed a pilot feasibility trial with a modest number (14) of stroke survivors, and compared it with an equal number of matched controls. Miraculously, we had done it in under a year and for less than $100,000. Unfortunately, however, we weren't anywhere near the hundreds of research participants that we needed to convince the FDA that we should be able to sell this product to stroke survivors and make the claim that we can improve motor function. Without a stamp of approval from the FDA and the ability to create billing codes from our product, clinicians in major rehabilitation institutes were not going to play ball in "prescribing" our software. End conclusion: we needed more money to get to where we needed to go. We applied for National Institutes of Health (NIH) grants to fund our clinical trials, but, in their infinite wisdom, our reviewers told us that our approach was not exciting or novel enough to warrant funding. I don't really blame the NIH, though: they are so chronically underfunded that their reviewers are basically searching for reasons *not* to fund the grants they have to review.

We met with Venture Capital (VC) firms—they were interested in the principle, but they either wanted a solution that was already far more clinically validated than what we had (why would we need the money, then?) or they wanted to have far more ownership of the company than was prudent for us to hand over. Although we don't go into too much detail about the financial aspects of running a company in this book, let's unpack this concept a little, because it is important for a lot of first-timers to hear. Obviously, all companies need funding in their developmental stages. Even if you are the leanest of startups working under the most Spartan of conditions, if your product starts to take off, you will need funding on hand to be able to successfully scale it to the masses. This is where investors come in. One of the most common ways of getting serious funding for one of your ideas is to get a VC involved. The term 'VC' gets thrown around a lot, let's see if we can demystify it. The role of a VC firm is to make strategic investments in promising young companies in return for ownership or equity of that company. The money allows the company to grow at a critical period, and the VC firm impatiently waits for revenues to start rolling in so that they can enjoy their Return on Investment (ROI). Probably the most widely known example of how VCs operate is the hit TV show 'Shark Tank'. If you've never watched, but you're interested in an entertaining crash course in understanding how a lot of VCs approach investment in companies, I'd highly recommend watching this show, where you get to watch a super-charged, highly accelerated version of the process take place. The basic process is as follows:

1. You have an idea and found a company. You own 100% of the company, but you need money.
2. You pitch your idea to any VC firm/private investor that will listen. Your job is to answer (with as much clarity as possible) some key business questions about your product:

   a. The potential market size
   b. Why your team is the best team in the world to make your idea happen (and better than any competitors already out there)?
   c. How quickly you can move from 'idea' to 'revenue' (how quickly your investors will see a ROI)?
   d. What share of the global market do you think you can realistically corner if you successfully scale?
   e. What is your long-term vision/exit strategy for your company (are you looking to be bought out by a larger company, or grow into one)?

3. If the VC likes your pitch, they will offer you funding in return for a percentage of your company. You should, of course, negotiate back and forth with them to make sure you get the best possible deal, but this is the step where you have to be really careful.
4. Take the money and run! Just kidding…*sigh*…take the money and grow your business into something profitable for your investors.

The reason that step 3 in the process is so critical, is because many companies give up too much equity in their company too early, which can place them in all sorts of difficult positions later down the track.

Ideally, you want to build your company to a point where it is profitable without the founding members having to give away the controlling share of the company to an outsider. This is because many outside investors may not have the same mission as you. For instance, let's say GesTherapy negotiated an agreement with a VC where they gave us $1,500,000 in return for 70% of the company. That 1.5 million sounds good, but the 70% control doesn't. The VCs own us now: they could have replaced everyone on our board with their own people, and edge us out of our own company. Maybe they don't even care about stroke rehab—maybe they saw something they liked in our software and decided to see if they could build that into something profitable. GesTherapy's mission to improve rehabilitation goes down the drain! Now, in the blink of an eye, I'm the part-owner of a company that I have no control over, that is doing something completely different to what I intended, but that also has my name all over it. Sure, we have 1.5 million in the bank, but I can't touch that money, and they just sacked me from any salaried role in the company. At this point, I can either try to cash out: sell off my remaining equity in the company (often very hard to do) and run, or wait around and hope that whatever the new majority owners of my company build actually turns out to be valuable. Either way, I'm not exactly living the dream of a successful entrepreneur building my dream product…and now, to cap it all off, the VC firm/evil investor owns my precious, beautiful idea.

The other thing to look out for is giving away too much equity too soon because you may need more equity later. For instance, let's say a VC offered GesTherapy $1,500,000 in return for 70% of the company, but we were way too savvy for that: we negotiated them down to 49% of the company. Now we still own the controlling 51% (assuming we don't devolve into infighting and have one of us defect). This may sound like an OK situation, but we had better make sure that 1.5 million is all we need to turn GesTherapy into a profitable telerehabilitation company (spoiler alert: it definitely isn't). So now let's pretend that we spend the 1.5 million to very good effect, but we're still not quite at the point where the company is profitable— we need another 1.5 million to get there. Now we're in a tough spot: we gave away too much equity in the first round, and now it is going to be very hard to raise the money we need without losing control of the company, and perhaps even losing so much of our own equity in the company that it is no longer a worthwhile venture for anyone. Long story short, be strategic when you're considering potential investors —the money may seem really great at first, but there are always strings attached, and the wrong investor can be incredibly predatory if you're not vigilant.

OK—moving back to our story's unhappy ending. Unfortunately, we simply couldn't find anyone to fund the trials that we needed to take our product to the next level. The company disbanded, and I was quite sad because I really felt as though something that had the potential to be transformative for the stroke recovery landscape had been let down by a lack of vision and/or unwillingness to spend investment money on carefully conducted trials. I chose to add this case study into the book because it taught me some incredibly important principles about the development of a digital health product. The first thing that really struck me about this experience was something that I have mentioned previously in this book:

developing a digital health product is so much more involved than building a tech product. If we were running GesTherapy like any other tech product, we would have just ridden the press wave that came about from our initial user trials, released our software online, charged a subscription and let the money roll in. But things aren't so simple. Let's pretend for just one moment that we were encouraged enough by the media attention and emails/calls from prospective customers that we decided to throw caution to the wind and release our product as some sort of generic "digital exercise therapy". We wouldn't make the 'Lumosity' error—we wouldn't tell people (or even remotely imply) that it is a product for stroke survivors, we would just say that it is an exercise product that seems to help some people keep or regain hand function, right? We'll let the media do the rest as our client base continues to grow…right? Probably not. Well, probably not in any ethical way when you get right down to it. The first issue that we have to face is that "one size fits all" medicine doesn't really work—especially when it comes to rehabilitation medicine. Different stroke survivors need different exercises to make important functional gains, so we need someone with domain knowledge (a physical therapist or occupational therapist) to prescribe appropriate exercises and make changes based on progress (or lack thereof). This is non-negotiable for effective therapy, and it leaves you with a choice to make about your product. Do you:

(a) Try to make a quick buck with a **tech** product that probably won't work for a majority of patients, but might work for *just* enough of your target population to allow your product to join the ranks of many other alternative therapies with questionable efficacy.
(b) Create a validated, FDA-approved platform that combines an innovative digital technology with skilled health professionals who can prescribe, monitor and progress therapeutic exercises with unprecedented accuracy.

   I know which one I would prefer, but it sure is hard to get off the ground (b. I choose option b). For the record, I still think that this is a good idea, and I hope that one day a group (maybe even my group) will be able to find the winning combination of clinical domain knowledge, clever software and hardware design, solidly funded clinical trials and the regulatory contacts to get a stroke rehab product developed, validated and approved to a point where physical therapists are allowed to bill for it. Since we're entering an aging crisis, this is a product that we desperately need, so I really hope we crack it one day soon.

Chapter 7
The Social Good Organization

"They didn't know it was impossible, so they did it."
—Mark Twain.

At the time of writing this, I have been volunteering for a group called "Not Impossible Labs" for about five years, and thanks to this group, I have one of my favorite professional titles: they call me their "Chief Mad Scientist". Beyond their penchant for bestowing ridiculous work titles, I also enjoy working with Not Impossible because they have managed to significantly impact various aspects of HealthTech in their short lifespan as a company, despite *not* being a traditional HealthTech company. Not Impossible is the brain child of Mick Ebeling and Elliot Kotek. Mick is a tall (many would say unnecessarily tall), bald, dynamo of a human being, with an annoying amount of energy, and an almost physical aversion to being told "no." You can read all about Mick's story in his own book *"Not Impossible: The art and joy of doing what couldn't be done."* Meanwhile, Elliot is the *Yin* to Mick's *Yang*—a wise, calm and kind individual who has an almost mystical ability to take in incredible amounts of data generated in chaotic, creative environments, and synthesize them into an actionable plan for production. Together, they set out to change the world, using technology for the sake of humanity.

As a company and a concept, Not Impossible's origins can be traced back to one person: Tony "Tempt One" Quan (Fig. 7.1). As a young man, Tempt was an up-and-coming American graffiti artist, whose unique, innovative style saw him gaining status and recognition within the Los Angeles and greater graffiti scene. But in 2003, Tempt was diagnosed with Amyotrophic Lateral Sclerosis (ALS). In case you've never dumped a bucket of ice water on your heads during the "Ice Bucket Challenge" viral sensation of 2014, ALS is a terrible, neurodegenerative condition with no cure. Cell by cell, the neurons in your body that are responsible for movement—your motor neurons—die off, leaving you a prisoner inside of your body. All of your senses work, your cognition is intact, but you have no ability to move, speak or influence your environment. Sometimes people with ALS or similar neurological conditions are aptly referred to as being "locked in". As far as tortures

© Springer International Publishing AG 2018
D. Putrino, *Hacking Health*, https://doi.org/10.1007/978-3-319-71619-0_7

Fig. 7.1 A self-portrait of Tony "Tempt" Quan (Image courtesy of Not Impossible Labs)

go, I feel as though we can go ahead and give Mother Nature the gold medal for thinking up one of the worst.

Over the course of the next few years, Tempt became confined to a hospital bed, he was intubated in order to help with his efforts to breathe, and he lost the ability to communicate independently. By 2009, the only way Tempt could communicate was to use a letter board—made famous in the 1997 book by Jean-Dominique Bauby, *The Diving Bell and the Butterfly*. Don't let the fancy name fool you, a letter board is just a sheet of paper with all the letters of the alphabet written on it. A caregiver/family member/health professional will point a finger at each letter on the board until they arrive at the one you want. Then you blink, and they write that letter down. Then they start the process again, and letter by letter, you painstakingly spell out the thing that you want to communicate to the person in the room.

As modes of communication go, this has to be one of the most inefficient ones around. It is incredibly frustrating and labor-intensive for everybody involved, which makes it more likely that a locked-in individual will find themselves socially isolated because few people want to regularly go through the incredible effort required for a regular conversation. But, it *does* happen to be free (if you have access to a pen and paper), which unfortunately makes it one of the most commonly used modes of communication. At the time, by contrast, eye-tracking technologies that would allow Tempt to communicate independently were available, but for around $40,000 a pop—not exactly accessible technology for an individual who is already struggling to scrape together the monthly costs of essentials like good nursing care. In short, Tempt was in a bit of a bind. Enter Mick Ebeling—not a health professional, entrepreneur, engineer or eye-tracking expert. Just a guy that didn't think eye tracking should cost so much. Mick's thought process wasn't more complicated than this: "why does an eye-tracker cost $40,000? Maybe it shouldn't." So he kicked the tires on this concept by bringing in a team of programmers and engineers and applying them to the problem. None of these guys were eye-tracking gurus, and objectively speaking they had no business being there. But if you put another lens on it, they were fresh to the problem and they theoretically had the skills to solve the problem. Mick gave them the same basic directive he gave himself: It seems crazy and unjust that eye-tracking technology costs so much. Let's make it not cost so much.

Mick's intuition proved to be correct, and within a few weeks this little group of hackers and makers had created the "Eyewriter," (Fig. 7.2) which is a piece of eye-tracking technology made from a cheap pair of sunglasses with the lenses popped out, some coat hangers, zip ties, duct tape, and a couple of webcams. The whole thing cost $20 to assemble and the software was open source.

They put the Eyewriter on Tempt and after training with the device, not only did he regain the ability to communicate independently, but, with the help of a special program that the team had made for him, he started to draw again. He regained his

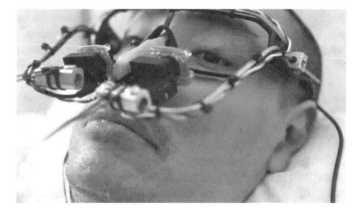

Fig. 7.2 Tempt wearing the Eyewriter (Image courtesy of Not Impossible Labs)

Fig. 7.3 Tempt using the
Eyewriter to make art for the
first time in 7 years. The
Eyewriter team used a laser to
sketch his designs on the wall
of a parking lot (Image
courtesy of Not Impossible
Labs)

ability to make art. The Eyewriter crew thought it would be cool to give Tempt the
ability to write graffiti again, so, one night, they hooked the Eyewriter up to a laser
that was pointed at a wall in a parking lot near the hospital where Tempt lives and
enabled him to control the laser via the Eyewriter (Fig. 7.3). A bunch of friends,
family, and admirers came by to watch him do his thing once more. I can only
guess how incredibly liberating and cathartic the experience must have been for
Tempt. The night was most definitely a success; everyone high-fived (probably),
had a beer (definitely), and considered the project a wrap.

What no one on the team expected was the degree of attention this humble
project brought in. The morning after their impromptu art show, the Eyewriter
started gaining attention…and I mean real attention: Time Magazine, Gizmodo,
TED, and so on. Within 6 months, the Eyewriter had clocked almost 300 million
media impressions and it was just getting started. It also got Mick thinking. With
his background in advertising, he is wired to assign great value to things that get
people clicking, watching, interacting, and engaging. The Eyewriter was an
intended operational success, but was also a completely unintended advertising
success, so perhaps this could be the basis of a new type of business for him.

Let's take a break here and discuss Not Impossible's first success, because I
think we can learn a lot from it. First of all, why did the project work so well? The
answer is that it had strong narrative and a clear mission. Every single person in the
room that was helping to develop the Eyewriter was doing it for Tempt. There was
no dead weight and everyone in the room had a focused, singular purpose. They
knew who they wanted to help, they knew what success looked like, they were
willing to work together to achieve it, and they were sufficiently naïve of the rules
in the industry, enabling them to bend them in order to achieve the seemingly
impossible. Some of this sounds a lot like what we've been talking about in this
book, but some of it is strikingly different as well. More on that later, but for now,
here are some of the process steps we learned from the development of the
Eyewriter:

1. *Identify the absurd*: In many industries, but particularly in the medical space, there are certain absurd things we take for granted. In this particular case, the fact that an eye-tracker cost $40,000 is f*cking absurd. I'm sorry to swear, but it is inexcusable. Especially when you consider the pain it causes so many people who need the technology, but can't afford it. To me, this step of the process doesn't require expertise or credentials, this is where you find your mission on a very human level. Find something that doesn't feel right to you on a basic logical or moral level, i.e. something that is causing people pain, and then you've found your mission.

2. *Find your 'One'*: Not Impossible has a mantra, which is: "Help One. Help Many." The basic idea here is that after you've found your mission, you find an individual that embodies that mission; the one person you want to help with your solution. Scalability comes after, first help your One. Now, as I mentioned in an earlier chapter, there are countless e-health companies that start their pitch with an anecdote: "my Grandma experienced x, and I vowed to myself, *never again*." We have already discussed some issues associated with this narrative, but another is that they often have an authenticity problem. I hear how Grandma inspired them to action, but then I never hear about Grandma again (spoiler alert: they're ignoring Grandma's calls so they can run their startup!). If your story starts with some "one," show me how you added value to their life with your invention, or you've lost me.

3. *The answer is always 'yes'*: It's amazing how liberating it can be to just say "yes" to something. In many scientific workplaces and communities the reflexive response to a new idea is to meet it with negativity. I don't think that this is because scientists are necessarily terrible people, I just think our training encourages us to push on a new idea from multiple directions and see how well it holds up to scrutiny. In a resource-scarce environment (and science is *very* scarce on resources!), this critical process can be helpful in de-risking new ventures. Thus, this cadre of "serious scientists" will never entertain a crazy idea or impractical request for very long. In return, they will likely enjoy long, successful and risk-averse careers, but their probability of truly *disrupting* their field is very low. In stark contrast, the Not Impossible approach is to say yes. No matter how crazy the notion, no matter how improbable the chance of success, if it is going to help our "one", then the answer is yes. This is the part of the process where we *commit*. We break the project down, start building teams, making calls, starting emails with "hope you're well…*now let me ask you for a massive favor*," generally drawing up some sort of actionable plan to get to the finish line.

4. *Act*: To borrow from a tiny little brand (I'm sure they won't mind): Just. Do. It. Not Impossible's process leans very heavily into the *"ask for forgiveness, not for permission"* mentality. I want to point out that this is not, nor should it be, the mentality of a HealthTech company. For all of the reasons we've discussed in previous chapters, building a HealthTech product should be a carefully cultivated process. One of the reasons I love volunteering for Not Impossible is getting to work on things that catalyze change on a timeline that feels like the

speed of light from my perspective. I'm used to working at the "scientific/ clinical trial pace." This does not make Not Impossible projects better or more important than the science I work on, it is just gratifying to me and my process to work on something that takes weeks to months instead of years to decades.

I first met Mick when Not Impossible was still reeling from the Eyewriter attention. I was looped into their next ambitious project called the Brainwriter, which is an affordable generation of devices that gives people the ability to communicate with their thoughts, even when eye movement has been lost. My team and I were on the project for a couple of weeks and Mick was a shadowy figure in a couple of the Skype meetings. Early one morning around 5 am, I was emailing the Brainwriter group and I received an immediate email back from Mick. Turns out he was in NYC, having breakfast not 2 blocks from my apartment. I popped down to join him and we finally got to meet in person. I started to chat to him about Brainwriter business, but within about 2 s flat I realized that he was excited about something else. That something else was Project Daniel. Someone sent Mick a Time Magazine article, which was written by Alex Perry in April of 2012, about an American doctor named Dr. Tom Catena's work at Mother of Mercy Hospital in South Sudan. Dr. Tom is an amazing human who is single-handedly keeping a hospital afloat in South Sudan (Fig. 7.4). Despite having no money, no supplies, no aid, and being right in the middle of an active war zone (the hospital itself has been bombed by President Omar al-Bashir multiple times), Dr. Tom does what he can. Much of the bleak story focused on a young boy named Daniel, then 14, who had lost both of his arms in a bombing. Despite having his life saved and residual limbs mended by Dr. Tom, Daniel did not think too much of his future. The author of the piece ends it with a heart-breaking sign off:

> *"Without hands, I can't do anything," says Daniel. "I can't even fight. I'm going to make such hard work for my family in the future." He looks me straight in the eye. "If I could have died, I would have," he says.*

Well if those last few sentences weren't a red rag to a bull-headed Mick Ebeling, I don't know what is. He jumped from step 1 in the playbook to step 4 in record

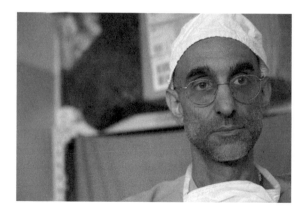

Fig. 7.4 After all international aid organizations had abandoned the location, Dr Tom Catena was the lone physician working at the Mother of Mercy Hospital in South Sudan (Image courtesy of Not Impossible Labs)

time. He knew what he needed to do: build Daniel some arms. So there I sat, probably having been awake for 30 min, with Mick laying out his strategy to help Daniel. I listened and asked him if a PT or anyone with rehab experience was going to be on the team. "No…do you think we need someone like that? Do you know anyone like that?" This was about the point where I realized that Mick really only knew me as David the neuroscientist (who lives in Boston), not David the PT, PhD who lives in NY. I told him what I do, his eyes widened a little, and then he got on his phone and pulled out his credit card. I asked him what he was doing—he looked at me like I was a complete idiot—"booking you a flight to LA". The next couple of hours were a bit of a blur. Mick couldn't book me on the same flight as him, so we went to the airport together, he marched me to the American Airlines counter, pulled out his iPad and showed the poor lady behind the counter photo after photo of African refugees in need until she broke down and moved my reservation so that we were on the same flight free of charge. We then worked tirelessly with an amazing team of people and (somewhat miraculously) around 2 months after my kidnapping, Mick was in South Sudan. He found Daniel in the largest refugee camp in the world, and brought him an upper limb prosthetic (Fig. 7.5a) that allowed him to feed himself for the first time in 2 years (Fig. 7.5b). If that wasn't sufficiently mythic, he went on to set up the world's first self-sustaining, community-run, 3D-printed prosthetics clinic in Dr. Tom's Hospital.

We didn't really get that much press, though.

JUST KIDDING! It clocked around a billion media impressions in 6 months, won every innovation and advertising award you can think of, and inspired millions worldwide.

After Project Daniel, Not Impossible has continued its good work and engaged in many other projects globally with a similar trajectory. The organization continues to be a force for global, social good and I'm proud of what we have achieved together. Interestingly, though, whenever I present a Not Impossible project at conferences, people are confused about how to frame the company. Invariably, they end up asking me questions like: "how many Eyewriters have you sold?" or "are your prosthetics FDA approved?" At first, I found these questions to be really peculiar, but eventually I realized that the people asking me these questions were wannabe e-health entrepreneurs who thought I found the Holy Grail: how to save the world and make a fortune along the way. But in reality, this isn't the case.

Not Impossible works project-to-project by finding an organization willing to sponsor each initiative in return for access to the media content we generate. The solutions that come out of Not Impossible are never intended to be brought to market and make a zillion dollars—that isn't the goal. Not Impossible, traditionally, has worked as a **disruptor** in the health technology space. With this in mind, let me introduce the fifth and final step in the Not Impossible process: "*Permission.*"

Before the Eyewriter, our friends making the $40,000 eye tracker were the only show in town. Shortly after the Eyewriter media hit it big, at least 5 companies making eye-tracking solutions for individuals with ALS (or conditions that have similar needs) emerged. Similarly, pre-Project Daniel, there weren't a whole lot of open source, 3D-printable prosthetics available online. These days, you can't swing

Fig. 7.5 a Daniel showing off his brand-new arm (Image courtesy of Not Impossible Labs).
b Daniel using his prosthesis to feed himself for the first time in 2 years (Image courtesy of Not
Impossible Labs)

a digital cat on the internet without hitting a new design. Whenever you achieve
something disruptive, you open up a new real estate in the environment you disrupt
and give others *permission* to enter that space. We've met many companies making
devices FAR better than our rough little prototypes that tell us they entered the
market after watching one of our videos. It's incredibly validating and humbling to
say the least.

   The very important point to take away from this case study is that Not
Impossible, for all its accolades, success and social good, is NOT a HealthTech
company. It is not run like a HealthTech company, and if they tried to be a

HealthTech company, they'd likely be sued and shut down in pretty short order. This is an important distinction to note: Not Impossible currently engages in *innovative disruption*, which is a vastly different process, and hard to get right and do responsibly. Much of my role as a volunteer at Not Impossible is to guide a lot of the science behind the stories that they tell. I work to ensure that the problems that are solved in the inspirational video content that Not Impossible produces are actually being solved in an authentic way. In considering the burgeoning HealthTech field, Mick and I often have conversations about how best to leverage some of the work that gets done at Not Impossible into an "incubator" of sorts—spinning off viable, ethical HealthTech companies that amplify the good they do. I'll let you know if they ever work it out.

Chapter 8
The Human Performance Division

"Baseball is 90 percent mental and the other half is physical"
—Yogi Berra

In our final case study, we're going to look at HealthTech through a different lens: that of human performance. For the last four years or so, I've been working as a collaborator with Red Bull's human performance division. It all started out innocently enough: one day, a good friend and colleague of mine, Dr Dylan Edwards, invited me to join him at an event in New York that was being held at the Australian Consulate. Turns out, a legendary Aussie soccer player, Tim Cahill, was giving a speech there. I'm not much of a sports guy, but they DID say there would be free booze at the event and my evening was open so I decided to tag along. Cahill ended up giving an amazing talk—thoughtful, reflective and honest, which I really enjoyed.

After the talk, there was an opportunity to meet-and-greet with Cahill, and you know how it is with soccer…people lost their damn minds. Dylan streaked ahead to get a photo with Tim for his son. I didn't really want to tangle with the crowd, so I hung back with my beer and saw that there was only one other person doing the same. I wandered over to her and we observed the fray together. "Don't you want to meet Tim?" she asked, "Nah. I'm not really that into soccer, but I loved his talk. What about you?" she shrugged, "I chat with Tim all the time, I'm his *can't really remember the phrase she used here, but I think it was something like *handler*."

So, we got to talking and it turns out Tim was a Red Bull-sponsored athlete (not a huge surprise given that, at the time, he was the captain of the New York Red Bulls). The other thing I learned was a little more surprising: Red Bull is seriously interested in the field of human performance. I really loved this idea and to me, this was like Amazon starting off by selling a physical product and then revolutionizing their field by becoming a repository of online knowledge/data (Amazon Web Services). It was cool, and I wanted in.

I spent some more time chatting with this lady at the consulate and she introduced us to Andy Walshe, who was the Director of the Red Bull High Performance program. Andy is an amazing individual who really exploded on the scene in 2012 when he helped to coach Felix Baumgartner to his stunning (and successful!) jump

© Springer International Publishing AG 2018
D. Putrino, *Hacking Health*, https://doi.org/10.1007/978-3-319-71619-0_8

to earth from the stratosphere. If you've never watched it, put this book down, go to YouTube, type "Red Bull Stratos" into the search bar, and get your mind blown. At the time that it happened, it took the title for the "most watched thing on the internet ever." It was a really incredible project and Andy learned a lot of things from it. While it's really Andy's story to tell (and he tells it very well), my abridged version is that mounting an endeavor like Stratos is not just an engineering problem, it is an athlete preparedness puzzle: keeping their body fit, keeping them physically and mentally engaged even during periods of exhaustion, frustration or boredom, keeping relations between the support team (that is equally responsible for success of the mission) and the athlete mutually respectful, productive and focused on the goal at hand, etc.

After meeting Andy for only a few minutes and hearing him talk passionately about these issues, I wasn't thinking about high performance so much, I was thinking about rehabilitation medicine. I started to understand that all patients are individuals trying to boost their performance, they're just starting out from a place further down the line than an elite athlete. It became clear to me that we had so much to learn from the team of human performance experts that Andy assembled, and I became passionate about being involved in his program.

We were basically falling over ourselves to work with Andy at this point, so we made the big ask: what could a couple of physical therapist/neuroscience/physiologist/scientist nerds do to help his mission? He shared with us that one thing he found really fascinating was the rapidly emerging field of transcranial Direct Current Stimulation (tDCS). Now, I recognize that tDCS may not be a household name for all of you, so let's do a crash course. A little while ago, some people noticed that if you took a positive electrode (red jumper-cable) and a negative electrode (black jumper-cable), attached these electrodes to highly specific points on your head, and plugged a battery into the circuit (direct current), that little current can make it past the scalp, skull, and all the other tissue and actually electrically stimulate brain tissue (Fig. 8.1a). Crazy, right? This observation has created somewhat of a fad industry, with people claiming you can create DIY tDCS devices that make you smarter/better/faster/stronger. I, for one, find it to be a sign of intelligence to *not* strap home-made electrical stimulation devices to my head, but maybe that's just me. There are also ready-made consumer electronic devices that you can buy online which claim to *maybe* make you smarter/better/faster/stronger (Fig. 8.1b). No one has ever proven these claims to any level of scientific rigor or satisfaction, though, so hang tight. The human brain stimulation revolution is not as close as you might think, in spite of all the over-hyped media articles about it.

In light of all of this, you may find yourself wondering what tDCS is *actually* good for. Well, the people doing the careful science have a few theories. I'm not going to delve into the details too deeply, but there are many people who are currently investigating the application of tDCS for treatment of a wide variety of conditions—everything from stroke to treatment resistant depression. The guys at Red Bull had zeroed in on research conducted by Dylan, myself, and another wonderful colleague of ours named Mar Cortes. This research is trying to identify

(a)

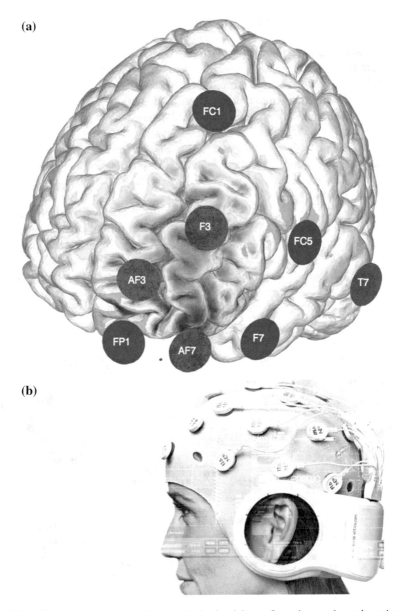

(b)

Fig. 8.1 a When you place a set of electrodes in the right configuration on the scalp and run a direct current through them, you can physically change the excitability of the brain (Image courtesy of Neuroelectrics). **b** An example of a wearable tDCS headset, developed by a rigorous and reputable neuroscience research company, that can potentially influence brain activity when used appropriately (Image courtesy of Neuroelectrics)

whether applying tDCS to stroke and spinal cord injury survivors prior to a neu-
rorehabilitation session can enhance the effects of the physical rehabilitation. We've
had some early results that might support this position. Andy was curious to know
if the same was true in athletes: if we push them to the absolute limit in training, can
we use tDCS to get just a little more out of them?

This was the question we sought to answer in our very first Red Bull experiment,
which was titled "Project Endurance 2.0." Endurance 2.0 was a massive effort and
you can watch how it all played out on YouTube if you search "Red Bull Project
Endurance 2.0." I wish I could name all of the amazing performance coaches,
physiologists and engineers we worked with on this project, but there's no space for
it. Suffice to say, Dylan, Mar, and I were humbled just to be in the room since every
single person working on the project was some version of a world-renowned,
international guru in their respective corner of the sports performance universe.
I don't want to speak for Dylan and Mar here, but I can tell you that my deeply
ingrained academic sense of "imposter syndrome" was tingling pretty intensely.
Then we had the athletes and I'm going to go ahead and name them: Rebecca
Rusch (Fig. 8.2a; aka "the Queen of Pain" who, incidentally, wrote an amazing
book called *"Rusch to Glory"* and starred in a stunning film called *"Blood Road"* if
you ever want the opportunity to get inside the head of an elite athlete for a little

Fig. 8.2 a Jesse Thomas (Image courtesy of Michael Darter/Red Bull Content Pool). **b** Tim
Johnston (Image courtesy of Michael Darter/Red Bull Content Pool). **c** Rebecca Rusch (Image
courtesy of Michael Darter/Red Bull Content Pool). **d** Mikey Day (Image courtesy of Michael
Darter/Red Bull Content Pool)

(a)   **(b)**

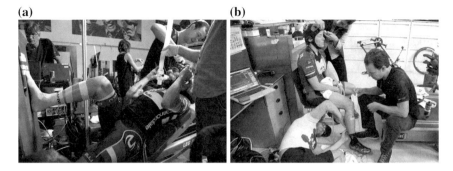

Fig. 8.3 **a** My colleague Dylan Edwards zapping Tim's brain with a massive magnetic pulse while he performs a maximal muscle contraction so that we can test the limits of his muscle endurance (Image courtesy of Michael Darter/Red Bull Content Pool). **b** Dylan Edwards, Mar Cortes and myself strapping electrodes to every inch of Rebecca's body so we can measure her muscle and brain responses during maximal exercise (Image courtesy of Michael Darter/Red Bull Content Pool)

while), Tim Johnson (Fig. 8.2b), Jesse Thomas (Fig. 8.2c), and Mikey Day (Fig. 8.2d). I feel the need to name these individuals because we worked them **so hard** and even while we were doing it, they were amongst the most gracious, professional, resilient, and unflappable human beings I've ever worked with.

Our job was to take these athletes (who were all world champions in their respective events), push them to the absolute limit (to muscle failure) and then ask them to go a little further (Fig. 8.3a). Then we did that for a bunch of days in a row. They had wires and devices hanging from every inch of their bodies as we recorded every biometric we could get our hands on, and some days we gave them real tDCS, other days we gave them a 'sham' tDCS condition (Fig. 8.3b; a stimulation protocol that makes it feel like you're getting your brain stimulated, but no current actually makes it to the brain). After all the dust had settled, we could confirm that tDCS had not, in fact, improved athletic performance in our brave cohort. It was a failed experiment. As with all good experiments, however, we had learned so much from the process, not just from an experimental design point-of-view, but also we learned some really interesting things about the neurophysiological control of a muscle that is just about to fail.

From there, we've had the opportunity to work with Red Bull on a bunch of other great projects. We were invited back for Red Bull 3.0, where we studied muscle physiology in athletes performing insane feats of endurance at different altitudes: from sea level to 6000 feet above sea level in the same day (courtesy of a little Cessna that flew the athletes from Death Valley to the top of Mammoth Mountain in California). In Salina Cruz, Mexico, we took to the water for Red Bull's "Surf Science", and attempted to study brain activation and responses to immersive virtual reality in some pro-surfers in an effort to understand how in the hell these guys are so good at what they do. That was one of the most fun/ frustrating projects we worked on because we had to learn how to waterproof all of our equipment before putting it on the surfers! We embedded a ton of stuff in silicone and we broke a lot of gear, but we had a great time doing it. In short, working with Red Bull allows us to apply bleeding-edge technology to the complex puzzle of how to move athletes further along the spectrum of human performance.

Why am I telling you all of this? Some of you may have guessed already. As I said in the beginning of this chapter, I firmly believe that there is a fine line between human performance work and clinical specialties such as cardiology, neurology and rehabilitation medicine. Performance coaching is all about taking a group of individuals that, on most objective scales, are as good as they possibly can be and discovering new and exciting ways to make them better. One of the greatest challenges we currently face in rehabilitation medicine is developing strategies and technologies to help a patient who has seemingly plateaued (according to our clinical scales) and enhancing their performance. The only difference between performance and rehabilitation medicine is that there is more investment and less budgetary constraints in performance coaching than in rehabilitation medicine. So what does this all mean to a group of budding HealthTech entrepreneurs? Your product may have its first start in the world of performance coaching before it makes it into medicine.

Sometimes, even if your device outperforms everything in the health ecosystem in allowing clinicians to characterize one aspect of their patients' physiology, it still isn't valuable to healthcare because it doesn't provide actionable data. Let me give you an example: Once upon a time, back when my body allowed it, I was on a field somewhere in the middle of Massachusetts playing a game of Australian football. Australian football (not soccer, not rugby) is a unique game that most Australians grow up playing. It can best be described as a cross between American football (the ball looks similar), lacrosse (the gameplay looks similar), and getting mugged (you don't wear pads/helmet and people can legally hit you from any direction). So we were mid-game and one of my teammates goes down with what looks like a sprained ankle. I immediately put my physiotherapist cap on and evaluated him. When it comes to ankles, there is a set of rules called the "Ottawa Ankle Rules," which are incredibly helpful guidelines in helping to decide whether or not the person you're looking at might need an x-ray versus 'you've just got a sprain, suck it up.' My friend was right on the border in this case. I wasn't happy, so we bundled him up and took him to the local hospital.

The doctor assessed him and sent him away for an X-ray, which turned out to be clear. We were gathering up his things to go, when our doc stopped us,

"Whoa whoa whoa! He still needs to go for an MRI!"

I was skeptical.

"Really? How much will that cost? He's visiting the US and doesn't have insurance."

"A few thousand dollars" he replied nonchalantly, as though that was a reasonable cost. I asked the doc, who was growing increasingly impatient with us, why exactly my friend required an MRI. He drew himself up, put on his best stern don't-argue-with-me-I-went-to-medical-school affect, and said:

"I need to determine the degree to which his anterior talo-fibular ligament has been damaged. It could be a grade II or a grade III tear."

Around this point I realized that our good doctor had assumed the mud-covered guy in shorts and a football jersey standing in front of him had zero medical knowledge. I internally (and maybe externally) eye-rolled, before replying.

"I'll make you a deal, doc: if you can explain to me right now how your medical management is going to change based on whether my friend has a Grade II or Grade III ATFL sprain, I'll pay for the MRI on the spot."

long pause

"OK. I'm discharging you." Yeah. That's what I thought.

The point is this story is that as we're rapidly moving toward models of 'accountable care' in healthcare systems all around the world, healthcare providers are only interested in new technologies that result in actionable changes in care-delivery. If a patient's outcome or management is not going to change (due to limited resources, or any other reason) based on how well you can sense the extent of their problem with a new device, no one is going to buy your device, even if it works perfectly! However, this is where renewed opportunity lies in the world of elite sports performance: if we're back in the emergency room, but rather than my friend visiting from Australia, I wander in with tennis superstar Venus Williams, you better believe she's getting that MRI, and ABSOLUTELY her management of the sprain is going to change based on the grade of the injury, which fibers of the ATFL were torn, and when her next grand slam event is scheduled.

The performance coaching field is complex and multi-factorial, but much of it lives and dies in our ability to sensitively monitor slight modulations in physiology and task performance. This is an incredibly feasible alternate pathway to commercialization that most HealthTech entrepreneurs should consider before they launch headlong into the healthcare industry.

Epilogue

"Success isn't about how much money you make, it's about the difference you make in people's lives."

—Michelle Obama

As we reach the end of the book, it's important to reflect on some of the major issues and lessons that surround the development of a HealthTech product. If you take away one message from this book, let it be this:

The development of a HealthTech product is not the same as developing a consumer technology product for the general population. It is a unique process that requires a deep level of healthcare domain knowledge and a team with a compassionate, humanitarian mission in mind.

This is the big idea, and the reason for this book. A unique process requires targeted literature to guide one through that process. I hope that this book is the first of many to form a rich body of literature around this crucial topic, which is sure to disrupt and reshape healthcare delivery in the 21st Century.

I've tried to make this book as readable and entertaining as possible, because we all know that reading textbooks can be boring. However, while I always heartily endorse having fun while you work, please don't confuse any playfulness on my part with glibness. Developing a HealthTech product is a very serious undertaking on a humanitarian level. If you get it right, many people are going to use your product to maintain their wellbeing, and maybe even depend on it to stay alive. This is a massive responsibility, and it shouldn't be taken on lightly. Rather, it should be approached with compassion, authenticity and a strong desire to solve a healthcare-related problem. This leads me to my second takeaway:

If you're just in it for the money, then you're in the wrong place.

It's true that healthcare is an emerging market on the technology landscape, and that there is tremendous potential and growth in this marketplace. However, HealthTech is a tightly regulated space, with many unique implementation and distribution challenges that other fields simply do not face. Success requires a lot of passion, perseverance and hard work, and even then, many successful HealthTech

© Springer International Publishing AG 2018
D. Putrino, *Hacking Health*, https://doi.org/10.1007/978-3-319-71619-0

products may not see a return on investment on the scale enjoyed by technology products in less regulated and specialized markets. HealthTech entrepreneurs must be willing to prioritize healthcare quality first and money second. If this doesn't sound like your sort of prioritizing (which is fine, by the way—the world needs all forms of technology innovation!), then make room for the possibility that the HealthTech world is not for you.

We've spent some time covering the need for an interdisciplinary approach to HealthTech and a diverse range of case studies. It is crucial for all HealthTech entrepreneurs to embed this sort of thinking into their world view. With this in mind, my final takeaway is:

Be interdisciplinary. Regularly look outside of your field and learn from the successes and failures of others.

You are entering a new and exciting market, and the fastest way to innovate in this space is to take ideas and wisdom from other fields. To paraphrase the great Steve Jobs, innovation is just mixing stuff together. It is important to point out that this advice does not conflict with the first takeaway. While you must remain mindful of the fact that you're in a unique field and that not all principles are generalizable, you can still enrich your product with wisdom from other fields. To illustrate this more concretely, let's take a moment to reconsider the trusty TKR recovery app that has been developed over the course of the book. Although this is a HealthTech product, you would be remiss not to study key successes and failures in the field of consumer app development for seniors before developing your app. Pairing TKR experts with app developers that are specialized to the senior market is a simple example of how an interdisciplinary approach will benefit product design, but through the case studies detailed in the later chapters I also hope you learned that you can get a little more creative with it. For instance, is Not Impossible's approach to story-telling the ultimate way to advertise a HealthTech product once it is well-validated? Is the (relatively!) risk-friendly sports perfor- mance market the ideal first client to create a sustainable revenue stream for your product before you approach the risk-averse health industry? Collaborate, learn from others, and spend some time outside of your comfort zone, because it will vastly improve your product.

Late at night, still on a high after drawing independently for the first time using the Eyewriter, Tempt wrote the team a message:

That was the first time I've drawn anything for 7 years. I feel like I had been held underwater, and someone finally reached down and pulled my head up so that I could take a breath.

Whenever I need a reminder of why I do what I do, I think about Tempt's message. It remains with me as a powerful example of how HealthTech can transform someone's life for the better. In many regards, we're pioneers exploring an exciting new field together, and it's hard to know what the rules are, so let's close out with some basics: solve problems that you care about, rigorously prove your effects with good science, and then find someone willing to pay for your product. If you can manage that, you'll be on track to save lives and make money.

Reading List

Throughout the book, I have promised you some reading lists on different topics. Here are a few books for each of those topics—Hiring, CEO advice, Healthcare Design, and Clinical Research.

Hiring Reading List

Good hiring is always going to loom large in any business that you build, and many would argue that regardless of your field, being effective and successful is entirely contingent on who you hire, so it seems like a good place to start doing a lot of reading. I selected a pretty diverse group of books that run the whole gauntlet from anecdotal and cultural books about hiring, to books that discuss good workplace culture, all the way to functional books about what to do in an interview.

Who: The A method for hiring,
Geoff Smart and Randy Street (Ballantine Books, 2008)

The no asshole rule: Building a civilized workplace and surviving one that isn't,
Robert I. Sutton (Business Plus/Hachette Book Group, 2010)

Work Rules! Insights from inside Google that will transform how you live and lead,
Laszlo Bock (Twelve, 2015)

Moneyball: The art of winning an unfair game,
Michael Lewis (W. W. Norton & Company, 2004)

Hire with your head: Using performance-based hiring to build great teams,
Lou Adler (Wiley, 2007)

96 great interview questions to ask before you hire,
Paul Falcone (AMACOM, 2008)

Keeping the millennials: Why companies are losing billions in turnover to this generation—and what to do about it,

© Springer International Publishing AG 2018
D. Putrino, *Hacking Health*, https://doi.org/10.1007/978-3-319-71619-0

Joanne G. Sujansky and Jan Ferri-Reed (Wiley, 2009)

First, break all the rules: What the world's greatest managers do differently,
Marcus Buckingham and Curt Coffman (Gallup Press/Simon & Schuster, 2016

CEO Reading List

I had a lot of help with pulling this list together, because a lot of my friends happen to
be, or work with, fantastic CEOs, and they all did it by reading and learning from the
right books. This list also ranges a couple of topics, from how to be a CEO, to how to
effectively manage people, and finally how to build a successful startup from the
ground up. I have to admit, that there are a couple of authors in here that have public
personas or world views that I don't necessarily like or agree with, but that does not
mean that we can't learn from them (and the content of the books is good)!

Good to Great: Why some companies make the leap and others don't,
Jim Collins (HarperBusiness, 2001)

The Effective Executive: The definitive guide to getting the right things done,
Peter F. Drucker (HarperBusiness, 2006)

Extreme Ownership: How U.S. Navy Seals lead and win,
Jocko Willink and Leif Babin (St. Martin's Press, 2015)

Superbosses: How exceptional leaders master the flow of talent,
Sydney Finkelstein (Portfolio/Penguin, 2016)

What got you here won't get you there: How successful people become even
more successful,
Marshall Goldsmith (Hachette Books, 2007)

Perennial seller: The art of making and marketing work that lasts,
Ryan Holiday (Portfolio/Penguin, 2017)

Start with why: How great leaders inspire everyone to take action,
Simon Sinek (Portfolio/Penguin, 2011)

Zero to one: Notes on startups, or how to build the future,
Peter Thiel (Crown Business, 2014)

Repeatability: Build enduring businesses for a world of constant change,
Chris Zook and James Allen (Harvard Business Review Press, 2012)

You're in charge–Now what?: The 8 point plan,
Thomas J. Neff and James M. Citrin (Crown Business, 2007)

The new leader's 100-day action plan: How to take charge, build your team, and
get immediate results,
George Bradt, Jayme Check and Jorge Pedraza (Wiley, 2016)

Healthcare Design Reading List:

Designing products for healthcare is most definitely its own thing, and should be treated as such. Design for health technology still has a long way to go, and to date I haven't been exposed to any truly great books that focus on this. Instead, I focused my search on books that explore healthcare design in the context of building spaces that provide healthcare and improve patient experience. The rationale here is that the basic, guiding principles explored in the books listed are solid, and generalizable enough to at least give you a peek inside the heads of people who think about design and healthcare all day long. You can make your own product-specific extrapolations from the foundation you have built from this reading list.

The healing of America: A global quest for better, cheaper, and fairer health care,
T. R. Reid (Penguin, 2010)

The healing of America: A global quest for better, cheaper, and fairer health care,
T. R. Reid (Penguin, 2010)

Value stream mapping: How to visualize work and align leadership for organizational transformation,
Karen Martin and Mike Osterling (McGraw-Hill Education, 2013)

The lean 3P advantage: A practitioner's guide to the Production Preparation Process,
Allan Coletta (CRC Press, 2012)

Transforming health care: Virginia mason medical center's pursuit of the perfect patient experience,
Charles Kenney (CRC Press, 2010)

On the mend: Revolutionizing healthcare to save lives and transform the industry,
John Toussaint and Roger Gerard (Lean Enterprise Institute, 2010)

Evidence-based design for healthcare facilities,
Cynthia McCullough (Sigma Theta Tau International, 2010)

Clinical Research Reading List

I'm not going to speculate as to why, but the world of clinical research is yet to produce any enjoyable books about clinical research. Hopefully someone will take care of that one day, but in the meantime here are some truly excellent books that dig into the details about clinical research. I have tried to range the reading list to cover topics that range from the basic principles of clinical research to how to conduct a clinical trial, to data analysis, management and presentation. Read these books and make sure you understand them. Your ability to successfully carry out the processes described in this reading list is what will separate the Tech entrepreneurs from the HealthTech entrepreneurs.

A clinical trials manual from the Duke clinical research institute: Lessons from a horse named Jim
Margaret B. Liu and Kate Davis (Wiley-Blackwell, 2010)

Pharmaceutical and biomedical project management in a changing global environment,
Scott D. Babler and Sean Ekins (Wiley, 2010)

Fundamentals of clinical trials,
Lawrence M. Friedman et al (Springer, 2015)

Designing clinical research,
Stephen B. Hulley et al (Lippincott Williams & Wilkins, 2013)

Publishing and presenting clinical research,
Warren S. Browner (Lippincott Williams & Wilkins, 2012)

Practical guide to clinical data management,
Susanne Prokscha (CRC Press, 2011)